Series on Special Topics
Volume 1

Sand Control

Sand Control

W.L. Penberthy Jr.
Senior Research Associate
Exxon Production Research Co.

C.M. Shaughnessy
Research Supervisor
Exxon Production Research Co.

Henry L. Doherty Memorial Fund of AIME

Society of Petroleum Engineers

Printed in the United States of America.

ISBN 978-1-55563-041-6

12 13 14 15 16 17 / 13 12 11 10 9 8 7

Society of Petroleum Engineers
222 Palisades Creek Drive
Richardson, TX 75080-2040 USA

http://store.spe.org/
books@spe.org
1.972.952.9393

Preface

This special-topics monograph is intended to serve as a guide and reference for assisting engineers with completions where the production of formation sand presents a problem. It contains details concerning the selection of the proper sand-control method, in addition to providing procedures for performing them. Experience has shown that adhering to the practices presented here will accomplish the primary completion objectives: sand control, productivity, and longevity.

The techniques and procedures discussed represent many years of research and field experience within Exxon Corp. and the industry. We are indebted to our associates at Exxon Production Research Co. for their many contributions: F.A. Brooks, B.J. Cope, E.E. Echols, J.W. Graham, G.A. Gabriel, C. Gruesbeck, G. Inamdar, L.O. Knott, T.W. Muecke, M.V. Rowe, W.M. Salathiel, C.O. Stokley, and M.E. Whiteley.

We have been guided by many excellent suggestions and comments by the Review Committee and other industry personnel. Their assistance is gratefully acknowledged, in addition to that provided by the SPE staff. Finally, we thank Exxon Production Research Co. for its encouragement, support, and permission to publish this book.

W.L. Penberthy Jr.
C.M. Shaughnessy
Feb. 1992

SPE Series on Special Topics

The Series on Special Topics was established in 1965 by action of the SPE Board of Directors. The Series is intended to provide state-of-the-art technology in selected fields. The Series is directed by the Society's Monograph Committee. A committee member designated as Editor provides technical evaluation with the aid of the Review Committee.

Monograph Committee (1991)

Acknowledgments

This book is the result of efforts of a group of very talented people, in addition to the authors. The manuscript was read and critiqued by individuals who are very experienced in the theory and application of sand-control practices. The people who helped review this book include Don Olivier, D.G. Olivier & Assocs. Inc.; David Norman, Dowell Schlumberger; Steve Shryock, Chevron Overseas Petroleum Inc.; Robert Torrest, Arizona State U.; and Jim Rike, Rike Service Inc. The help and dedication of these individuals are appreciated not only by the SPE, but by all who read and use this book.

David R. Underdown
Editor
March 1992

Contents

Chapter 1
Sand Production From Oil and Gas Wells

Introduction

The production of formation sand into a well is one of the oldest problems plaguing the oil and gas industry because of its adverse effects on well productivity and equipment. It is normally associated with shallow, geologically young formations that have little or no natural cementation to hold the individual sand grains together. As a result, when the wellbore pressure is lower than the reservoir pressure, drag forces are applied to the formation sands as a consequence of fluid production. If the formation's restraining forces are exceeded, sand will be drawn into the wellbore, as **Fig. 1** suggests. The sand either plugs the well or is produced.

Produced sand has essentially no economic value. On the contrary, formation sand not only can plug wells, but also can erode equipment and settle in surface vessels. Controlling formation sand is costly and usually involves either slowing the production rate or using gravel-packing or sand-consolidation techniques. Operators spend millions of dollars each year to prevent the production of formation sand and to deal with other sand-related problems. Expenditures of this magnitude obviously have a significant impact on profits. In spite of these costs, effective sand-control practices have yielded oil and gas from wells that otherwise would have been shut-in.

Causes of Sand Production

In considering sand control, or formation solids control, one must differentiate between load-bearing solids and the fine particles (fines) that are not usually considered a part of the mechanical structure of the formation. Some fines are probably always produced with the well fluids, which, in fact, is beneficial. If the fines move freely through the gravel pack, they will not plug it. Thus, "sand control" actually refers to the control of the load-bearing particles, those that support the overburden. The problem then is deciding what is excessive sand production, a decision that involves many factors. Sand production higher than 0.1% (volumetric) can usually be considered excessive, but depending on the circumstances, the practical limit could be much lower or higher.

Fluid Flow. The major stresses that tend to cause sand production usually result from fluid flow, which is proportional to the pressure drop between the wellbore and the reservoir. The drag force imparted by fluid flow is related to the product of fluid velocity and viscosity. Whether the sand is water- or oil-wet may also affect its tendency to be produced. Intergranular bonds provide the major restraining forces against these stresses. For this reason, a well's production rate and the degree of natural consolidation present in the formation are the chief determinants in sand production. Sand arches[1,2] may also form around perforations to reduce sand production. Experience has indicated that these arches usually are unstable and tend to degrade easily with time and changes in producing conditions. As a result, they tend to exist only temporarily but may re-form later, only to break down again as conditions change.

Geographic and Geologic Factors. Sand production has been experienced in almost every area in the world where oil or gas production occurs from sandstone reservoirs; thus, the problem is worldwide. Sand production is most common in Tertiary Age sand reservoirs. Because these reservoirs are geologically young and are usually located at relatively shallow depths, they are no more than moderately consolidated ($\pm$100-psi compressive strength). **Fig. 2** shows typical sieve analyses of sands from several geographic locations. Because older reservoir rocks tend to be better consolidated, sand production problems are not as severe. Nevertheless, depending on well rates, drag forces may be sufficiently high to cause formation sand production in reservoirs in which the formation's compressive strength is well over 1,000 psi. However, this is uncommon.

Production Rate. When fluids are produced from sandstone reservoirs, stresses are imposed on the sand grains that tend to move them into the wellbore along with the produced fluids. These stresses are caused by pressure differences in the formation, fluid frictional forces, and the weight of the overburden. When the combined strength of these stresses exceeds the strength of the formation, sand will be produced. The implication is that, for many wells, there is a threshold production rate (and associated pressure drawdown) below which sand will not be produced and above which the formation will fail. Unfortunately, this rate is often below the economical producing rate. In these situations, some means of sand control must be included as an initial or remedial part of the well completion. The mechanism that causes a consolidated sand to fail is believed to result from a combination of pressure drawdown and fluid flow. Because these mechanisms are closely tied to each other, determining the actual mechanism may be a moot point.

Natural Consolidation. Opposing the fluid forces are the restraining forces that act to hold sand grains in place. These forces arise from intergranular bonds (natural consolidation), intergranular friction,

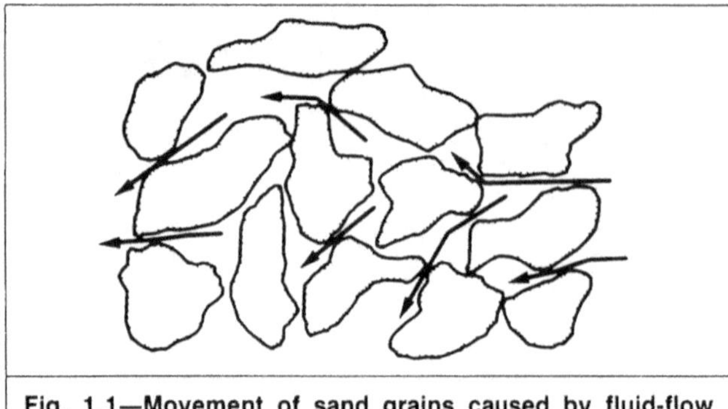

Fig. 1.1—Movement of sand grains caused by fluid-flow stress.

gravity forces, and capillary forces. Internal pore pressure (reservoir pressure) helps support the weight of the overburden, thereby acting to relieve some of the stress on the sand grains. Of these forces, the intergranular bonds are the most important factor in preventing sand production. The compressive strength of a formation sand is probably the best measure of the intergranular bond. If good completion and production practices are followed, formations with a compressive strength exceeding 1,000 psi will generally produce sand-free, as noted above. The exception is the case where the pressure drawdown around the well is high. If the pressure drawdown is low, however, sands with much lower compressive strengths may also produce sand-free.

Time Dependence. Predicting the sand-producing tendency of a formation is complicated, not only because data on the various factors to be considered are limited, but also because many factors change with time. Some examples of time-related changes follow.

- Decreasing reservoir pressure increases the overburden stress on the sand grains.
- Water production may dissolve natural cementing materials and weaken intergranular bonds or change the capillary pressure.
- Permeability reductions resulting from relative permeability effects or fines invasion, paraffin and asphaltene deposits, etc., may increase the stresses induced on the sand grains by fluid flow at a given rate.

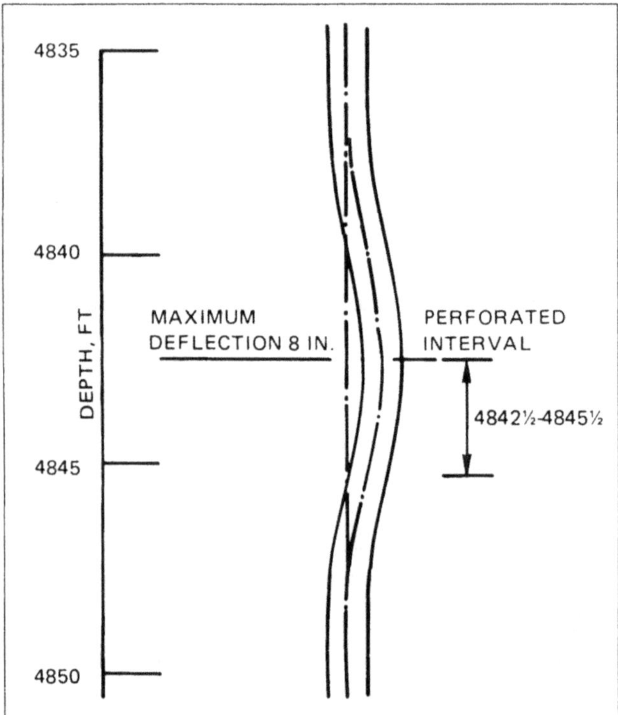

Fig. 1.3—Casing buckling caused by sand production in a well offshore Louisiana. The 7-in. casing was deflected 8 in. within a vertical distance of 5 to 10 ft.[5]

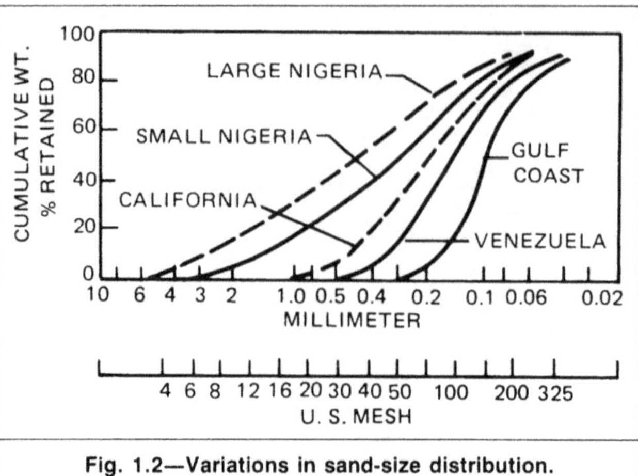

Fig. 1.2—Variations in sand-size distribution.

Because the problem is complex, the most meaningful prediction of potential sand problems is usually a correlation based on the performance of offset wells producing from the same reservoir.

Multiphase Flow. In many cases, sand production increases substantially when wells begin to produce water or gas with hydrocarbon fluids. Several theories have been given to explain this behavior.[3,4]

1. The produced water may dissolve part of the cementing material so that sand is produced.
2. The change in water saturation reduces capillary pressure (interfacial tension) to the point that the sand grains are not held together by these forces, causing excessive sand production.
3. When an interface moves through a pore space, a localized pressure disturbance is created. Water movement may also mobilize fines and ultimately cause plugging of the pore structure.

Fig. 1.4—Screen damage caused by sand production.

4. Relative permeability effects may also account for these sand-producing tendencies and may increase the pressure drawdown for a given well rate.

In most sand formations, the fine particles are water-wet and move with the water phase. Studies have clearly shown that multiphase flow can cause excessive sand production and mobilize the fines that create formation plugging, thereby reducing well productivity.

Thermal Effects. Thermal effects can also destroy intergranular bonds and influence sand production. The effects of high temperature associated with steam injection and combustion oil recovery projects have demonstrated that many wells experience high sand production not previously encountered in these operations.

Consequences of Sand Production

Several operational problems can arise if a well produces sand. All are troublesome and costly, but the degree of severity varies widely. Probably the least severe problems are solved simply by periodic removal of sand from such surface facilities as flowlines, manifolds, and separators. The more serious consequences relate to erosion of surface equipment, casing damage, and productivity losses.

Surface Equipment. Removal of sand from surface equipment is common in sand-producing regions. Perhaps only troublesome in land-based operations, sand disposal can become an acute problem offshore. Special treating facilities are required to remove the sand from the oil before offshore disposal. The problems arising from produced sand become more severe when the entrained sand is carried at velocities sufficient to erode surface equipment, such as valves and chokes, necessitating periodic replacement of these items. By far the worst complication of sand production is the erosion of surface equipment to the point that it fails, allowing high-pressure gas and/or oil to escape. This situation obviously constitutes a severe safety and pollution hazard. Equipment erosion by sand is much more severe when the well stream is in turbulent flow than in laminar flow.

Tubular Damage. Casing failure may accompany the production of formation sand in the producing interval, meaning the loss of a well. As sand is produced, slumping of the overlying casing-bearing formations can subject the entire casing string to abnormal loads. Such loads can lead to severe buckling when the lateral restraint provided by the surrounding sand is lost during sand production. In an example from offshore Louisiana, 7-in. casing was deflected 8 in. within a vertical distance of 5 to 10 ft (see **Fig. 3**).[5] Produced sand has also caused rod pumps to seize and rod strings to erode. Produced sand accelerates corrosion processes by exposing bare metal to the corrosive well fluids.

Produced sand can also damage downhole equipment, such as the wire-wrapped screen shown in **Fig. 4**.

Productivity Loss. Productivity is lost when a sand bridge forms in the production tubulars. This sanded-up condition occurs when the fluid velocity is insufficient to suspend the produced sand completely and flow it from the well. In settling out of the produced fluids, sand can then fill the production tubing and block the flow. The amount of sand fill sufficient to plug the tubing and to stop flow depends on well conditions. A simple Darcy flow calculation indicates that only a few inches of fill is required to reduce the flow rate substantially. These sand plugs must be removed by bailing or washing before production can be restored.

Effects of Sand Production on Formation

A question that generates considerable debate is, what happens to the formation when sand is produced? Some people believe that a cavity forms opposite the perforations. Others think that the sand, being essentially an infinite body, expands slightly with an infinitesimal increase in porosity to prevent cavities from forming. A third explanation is that the overburden subsides to fill any void regions created by sand production.

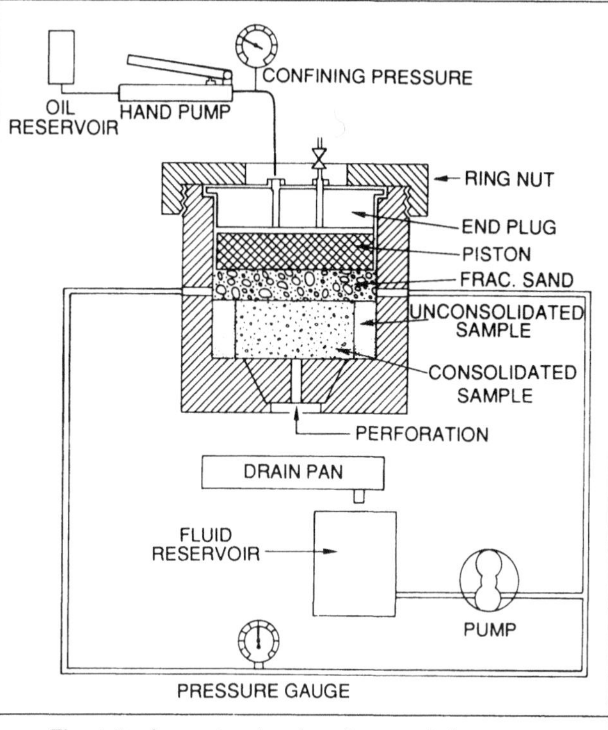

Fig. 1.5—Apparatus for drawdown-to-failure tests.

The answer to the above question is of considerable importance to both well performance and subsequent workover operations.

Experiments were conducted to characterize the behavior of sand when it fails under stresses induced by fluid flow.* These experiments were performed by placing 5-in.-diameter, 3-in.-long samples of consolidated sand in a high-pressure vessel. The bottom faces of the samples were placed next to a 0.5-in.-diameter hole, which simulated a perforation (see **Fig. 5**). Fluids were then flowed downward through the core at increasing flow rates (and pressure drops) until sand was produced through the perforation. **Fig. 6** shows samples removed from the test apparatus after varying amounts of sand had been produced. The early-time photograph shows that a cavity typical of a shear-type failure has formed, suggesting that the rock's compressive strength was exceeded. Two important conclusions were drawn from this study: fluid-flow stresses alone can cause sand production and cavities tend to form outside the perforation when sand is produced from formations with some degree of natural consolidation.

Sand-Control Techniques

Although the production and completion practices discussed later will reduce the frequency and severity of sand problems, many formations require additional control mechanisms.

Mechanical Retention. One obvious method of preventing formation sand from being produced is to restrain its entry into the wellbore flow stream. Many types of mechanical devices have been used for this purpose in the water-well and oil industries. Spherical particles will not flow continuously through rectangular slots twice as wide as the diameter of the particles, or through circular holes three times their size, as long as they flow in mass and have individual grain-to-grain contact.[6] The particles tend to "bridge" across such openings and prevent further particle movement. If the size of the particles varies, retention of the larger particles causes the smaller particles to bridge behind them.

All methods of mechanical retention are based on the principle of retaining a certain portion of the formation material to prevent the remainder from entering the well. Devices used by themselves to restrict sand movement usually are called screens or slotted liners. The placement of large, clean sand particles (actually called gravel

*Unpublished Exxon Production Research Co. reports, Houston (1973).

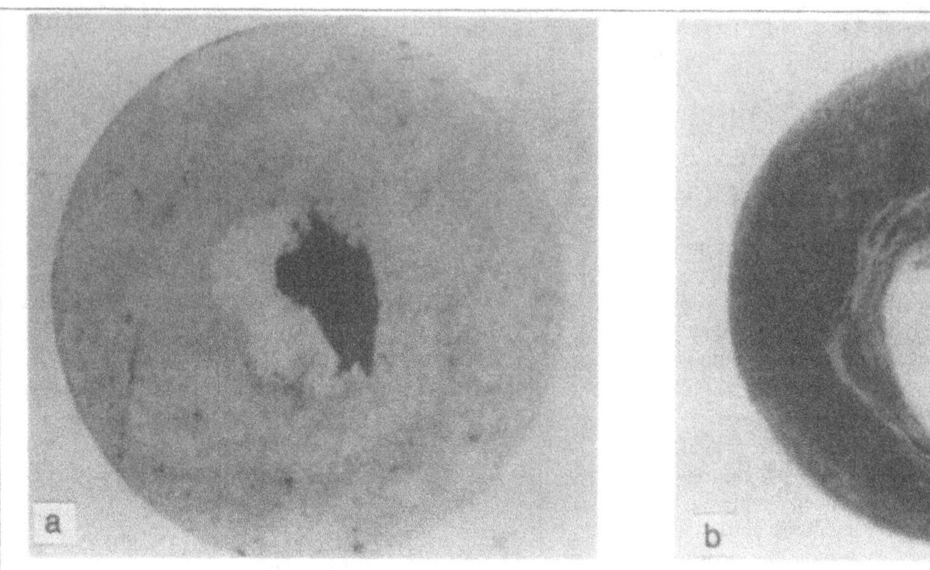

Fig. 1.6—Examples of cavity formation during drawdown-to-failure tests: (a) early results; (b) after completion of test.

because they are much larger than formation sand particles) between the screening device and the formation is referred to as gravel packing.

Regardless of which approach is selected, all mechanical sand-control methods are based on the bridging theory. The formation sand bridges against some sort of filtering medium, which allows the passage of fluid but prevents the production of formation sand. In addition to screens, other equipment used to control sand ranges from louvered casing to torch-cut and milled slots.[7] If only a slotted liner or screen is used to control formation sand, sand bridges should form against the slot when the slot opening is about 2.5 times the sand grain diameter.[8] Experiments have shown, however, that these bridges are not stable and that, for maximum effectiveness, the individual sand grains should bridge on, rather than in, the slot openings.

A screen or slotted liner run in a well to control formation sand usually is unsuccessful in controlling sand and/or maintaining the well's productivity for long periods of time. The exception to this is when the screen or liner is used where clean, relatively coarse-grained, high-permeability sand is present and a certain amount of formation sand production can be tolerated to produce the well economically.

The problem with screen-only completions relates to the manner in which oil and gas wells are completed with this technique. Here, the screen or slotted liner is simply run into position and the produced sand allowed to bridge around it. Because no back-surging or pump-in treatments are normally used (as is sometimes done in water-well completions) the finer particles are not effectively removed from the sandface of the near-wellbore sand mass. If these particles are retained on the screen or slotted liner, they can eventually plug it and substantially reduce well productivity. If they flow through the screen or slotted liner and do not plug it or form a sand plug in the well, the well may be produced indefinitely despite the fact that sand production has not been completely eliminated.

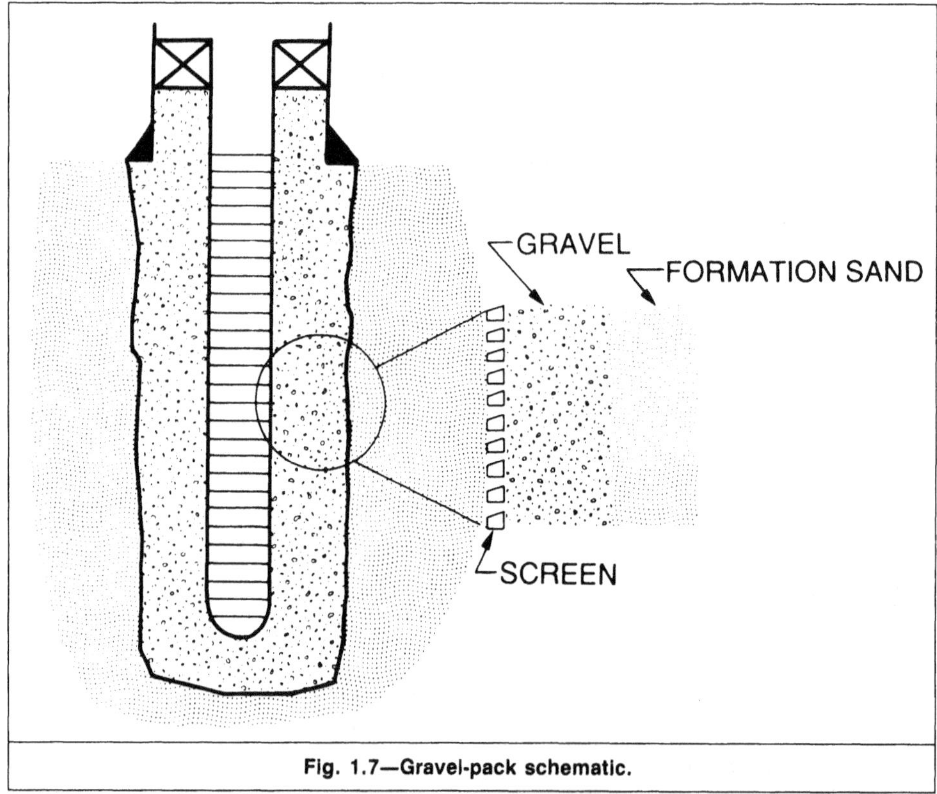

Fig. 1.7—Gravel-pack schematic.

A technique used on reasonably large-grained formation sand with limited success involves sizing the screen so that it will bridge particles larger than those at the 10th-percentile point on the respective sieve-analysis curve. This sizing allows the finer particles to be produced, resulting in a natural gravel pack formed from the larger sand grains. One problem with this approach is that the 10th-percentile point for many formation sands is smaller than 0.003 in., which is smaller than the openings on most commercial well screens. Consequently, it is not a practical application for many completions and is used only occasionally when gravel packing is not practical or when the formation sand is reasonably large.

Even though the screen/slotted-liner-only technique has been used to control sand temporarily, it is generally not recommended over gravel packing in terms of both productivity and sand control.

Gravel Packing. Gravel packing, the most popular mechanical sand-control technique, was used in the water-well industry long before it was used in oil or gas wells. It involves running a mechanical device such as a screen or slotted liner in the well and placing accurately sized gravel around the screen or slotted liner (**Fig. 7**). This placement allows the entry of fluids through the gravel but filters the formation sand from the flow stream so that sand-free production is possible. In most gravel packs, however, a finite amount of solids is produced, but they consist of the very fine particles that can move through the gravel pack. When performed properly, gravel packs yield long-life, high-productivity completions. Chaps. 3 through 10 contain details on gravel packing.

Chemical Sand Control. Chemical sand control (plastic consolidation) involves the use of plastics to cement the formation sand grains together artificially for several feet around the well so that formation fluids can be produced sand-free. To be effective, the plastics must (1) wet the sand surfaces and adhere to the sand grains, (2) yield a high compressive strength upon curing, and (3) maintain high well productivity.

In most field consolidation treatments, acid (mud acid), preflush (to remove residual water so that the plastic wets the sand grains), plastic resin, and then catalyst are pumped into the well. The well is then flushed and shut in to cure the plastic. The plastics usually are epoxies, furans, or phenolic resins. Chaps. 11 through 13 discuss the procedures used with the various consolidation processes.

Combination Sand-Control Techniques (Resin-Coated Gravel). Several combination sand-control techniques[9-15] use both gravel and plastic. The intent is to consolidate the gravel pack after it is placed but without the use of a screen or slotted liner. These processes are marketed under several trade names. All contain sand as the solid constituent and use epoxy, furan, or phenolic plastics.

The epoxy and furan techniques involve a resin-coated gravel mixed at the surface and pumped into the well. The gravel plastic slurry is then allowed to settle and cure. After curing, the residue is drilled out of the well before it is placed on production.

The phenolic resin-coated gravel processes involve a phenolic-coated gravel that is partially polymerized. Upon being subjected to temperatures higher than 135°F, the resin cure is completed so that the gravel is consolidated. Unlike the epoxy and furan processes, the phenolic resin-coated gravel is dry and can be handled much like ordinary gravel. Resin-coated gravel is discussed in more detail in Chap. 14.

References

1. Hall, C.D. Jr. and Harrisberger, W.H.: "Stability of Sand Arches: A Key to Sand Control," *JPT* (July 1970) 821-29.
2. Tippie, D.B. and Kohlhaas, C.A.: "Effect of Flow Rate on Stability of Unconsolidated Producing Sands," paper SPE 4533 presented at the 1973 SPE Annual Meeting, Las Vegas, Sept. 30-Oct. 3.
3. Muecke, T.W.: "Formation Fines and Factors Controlling Their Movement in Porous Media," *JPT* (Feb. 1979) 144-50.
4. Gabriel, G.A. and Inamdar, G.R.: "Experimental Investigation of Fine Migration in Porous Media," paper SPE 12168 presented at the 1983 SPE Annual Technical Conference and Exhibition, San Francisco, Oct. 5-8.
5. Suman, G.O. Jr.: "Casing Buckling in Producing Internals," *Pet. Eng.* (April 1974) 36-42.
6. Coberly, C.J. and Wagner, E.M.: "Some Considerations in the Selection and Installation of Gravel Packs for Oil Wells," *Pet. Tech.* (Aug. 1938) 1-20.
7. *Ground Water and Wells*, Johnson Div. UOP Inc., St. Paul, MN (1975) 151-52.
8. Coberly, C.J.: "Selection of Screen Openings for Unconsolidated Sands," *Drill. and Prod. Prac.* API (1937).
9. Sparlin, D.D.: "A New Sand Control Technique for Old Sand Problems," paper API 906-12-H presented at the 1967 Spring Meeting of the Southwestern District, Div. of Production, Midland, March 15-17.
10. "Here's a New Sand-Pack Method," *Oil & Gas J.* (March 10, 1969).
11. Copeland, C.T. and McAuley, J.D.: "Controlling Sand With an Epoxy-Coated, High-Solids-Content Gravel Slurry," *JPT* (Nov. 1974).
12. Sinclair, A.R. and Graham, J.W.: "An Effective Method of Sand Control," paper SPE 7004 presented at the 1978 SPE Symposium on Formation Damage Control, Lafayette, Feb. 15-16.
13. Knapp, R.H., Planty, R., and Voiland, E.J.: "A Gravel-Coating, Aqueous, Epoxy Emulsion System for Water-Based, Consolidated Gravel Packing: Development and Application," *JPT* (Nov. 1977) 1489-96; *Trans.*, AIME, **263**.
14. Constien, V.G. and Mayer, M.H.: "What? No Screen? Gravel Packing With Water-Carried Resin Coated Gravel," paper SPE 7003 presented at the 1978 SPE Symposium on Formation Damage Control, Lafayette, Feb. 15-16.
15. Suman, G.O., Ellis, R.C., and Snyder, R.E.: *Sand Control Handbook*, second edition, Gulf Publishing Co., Houston (1983) 67-68.

SI Metric Conversion Factors

ft	× 3.048*	E−01	=	m
°F	(°F−32)/1.8		=	°C
in.	× 2.54*	E+00	=	cm
psi	× 6.894 757	E+00	=	kPa

*Conversion factor is exact.

Chapter 2
Production and Completion Practices

Introduction

Production and completion practices can be used to combat sand problems as an alternative to gravel packing or sand consolidation. Both methods have been used alone or in combination to alleviate and/or control sand production. No one practice, however, can control sand effectively under all circumstances. Therefore, it is important that the advantages and limitations of each practice be recognized so that the best procedures can be followed to achieve high-productivity, sand-free completions.

Production Practices

Sand production is the natural consequence of fluid flow into a well. Therefore, reducing the production rate usually will diminish sand production as well. The tendency to produce sand is a reservoir characteristic. Some reservoirs can tolerate little, if any, fluid production without sanding up; others can sustain extremely high production rates with no sand problems. The sand-removal problem in some reservoirs is trivial and requires only periodic sand-removal washes. In other wells, sand may flow into the wellbore as quickly as it is removed, which can cause substantial lost revenues. The flow of formation sand into a well is related to the drag force created by the flowing fluid that occurs as a consequence of reducing the wellbore pressure below the reservoir pressure. Laboratory and field experiments have shown that sand production can be reduced by decreasing the production rate (drag force) to the point where the sand's natural cohesive forces hold it in place. Sand control, therefore, becomes possible through a step-by-step reduction in the production rate to a level at which the sand cut is acceptable. If the flow rate is high enough, no further sand control may be necessary.

Because sand production occurs when the stresses associated with fluid withdrawal exceed some threshold value, the probability of formation failure in any given well can be reduced by minimizing these stresses. Recognizing this fact, the oil industry in its early years controlled sand almost exclusively by limiting fluid production rates. This approach is still practiced; production from many wells is held below some assumed maximum rate thought to cause formation failure. Unfortunately, the maximum rate is difficult to predict accurately and to associate with incipient failure.

Unless the sand-free producing rate for a particular well remains high, any voluntary restriction of flow to prevent sanding represents a lost opportunity for maximizing revenue. In the past when wells were subject to proration, production from sand-prone wells was curtailed without a loss of profit. But the current demand for energy means that reduced production rates are equivalent to smaller profits. Reducing the production rate in these situations is not economically attractive, so other alternatives should be considered.

Sand Production Predictions

Although several investigators[1,2] have developed methods to predict sand production, no technique is universally accepted for predicting its onset accurately. Calculations and other analysis methods usually rely on a number of formation, fluid, and other properties that, in many cases, cannot be obtained from the rock formation with a high degree of accuracy. Even if techniques for acquiring the data were available, the process would be quite expensive. Experiments have shown that increased sand production is related to changes in production rate, water breakthrough, and multiphase production and reservoir depletion.[3-5] The interrelationship among all the variables, however, is not clear.

A few general correlations have been developed to help quantify this issue. These correlations are very approximate, however, and serve only as guidelines.

Correlations. Research has shown that sand production is related to the pressure drop and flow rate across the formation sand.* The investigation used the equipment illustrated in Fig. 1.5. Fig. 1.6 shows photographs of the cores removed from the equipment. The equipment was designed to determine the magnitude of the pressure drops that core samples could withstand before the onset of sand production. Tests demonstrated the following relationship:

$$\text{onset of sand production} \approx 1.7 \times \text{sand compressive strength},$$

where the compressive strength is in psi.

This correlation, intended to serve only as a guideline, indicates that sand production does not occur until a rock's compressive strength is substantially exceeded. The relationship applies only to consolidated sands; no correlation exists for unconsolidated sands. Assume that, in most cases, sand production will occur when fluid is produced. Other correlations are based on the Brinnel hardness number and the interval sonic travel time.

Log Analysis. Mechanical-properties log predictions[6,7] have been used to determine whether a formation will produce sand. The technique is based on calculations made from data taken from sonic, density, and neutron logs. While these predictions can determine

*Unpublished Exxon Production Research Co. report, Houston (1973).

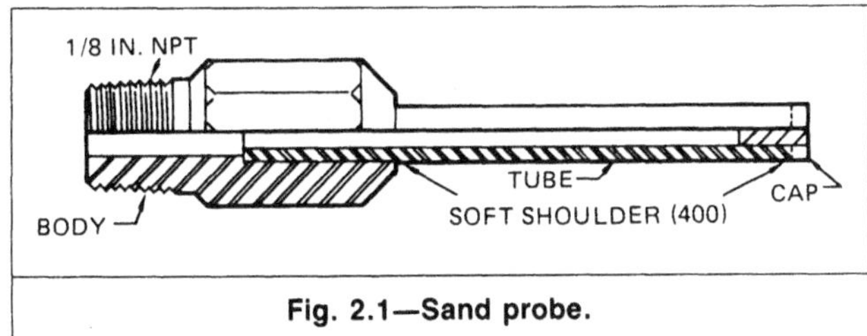

Fig. 2.1—Sand probe.

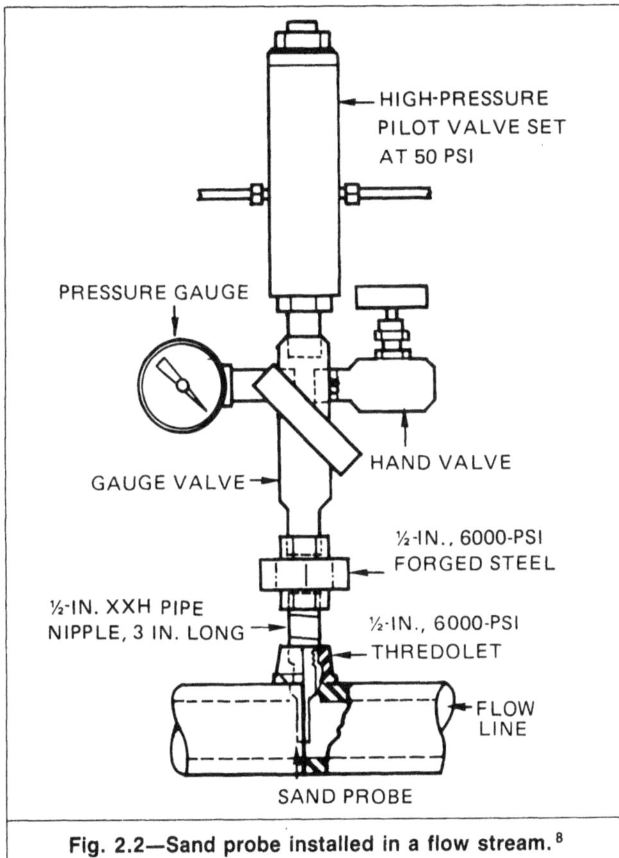

Fig. 2.2—Sand probe installed in a flow stream.[8]

those zones that are more (or less) consolidated than others, relying solely on this technique as a basis for determining whether sand control is necessary for a particular well involves many factors. A site-specific calibration of this technique is desirable before sand production can be predicted with a high degree of certainty.

The best method of determining whether a well will produce sand is to test it under field conditions.

Sand Detection

Sand-detection technology can assist in determining the maximum sand-free production rate. It also provides a method to monitor long-term success after application of a sand-control treatment. Devices to detect the presence of sand in produced fluids can substantially improve the safety and productivity of wells in sand-producing areas. Devices that detect sand produced to surface are sand probes[8] and sand detectors.[9,10] These devices enable the engineer to take remedial action before tubular goods and surface equipment are damaged, thereby increasing safety and decreasing workover costs. They can also be used to establish the maximum sand-free production rate for each well. Another device is designed to locate sand production through perforations. It also identifies damaged or corroded screens and casing within the wellbore.

Sand Probes. The sand probe (**Fig. 2.1**) is a hollow, stainless-steel cylinder sealed at one end. It is inserted into a flow stream with the open end protruding from the pipe wall (**Fig. 2.2**). When produced sand erodes the wall of the probe, flow-stream pressure is transmitted to a pilot valve, which closes the surface safety valve and shuts in the well. This equipment is applicable primarily to flowing wells.

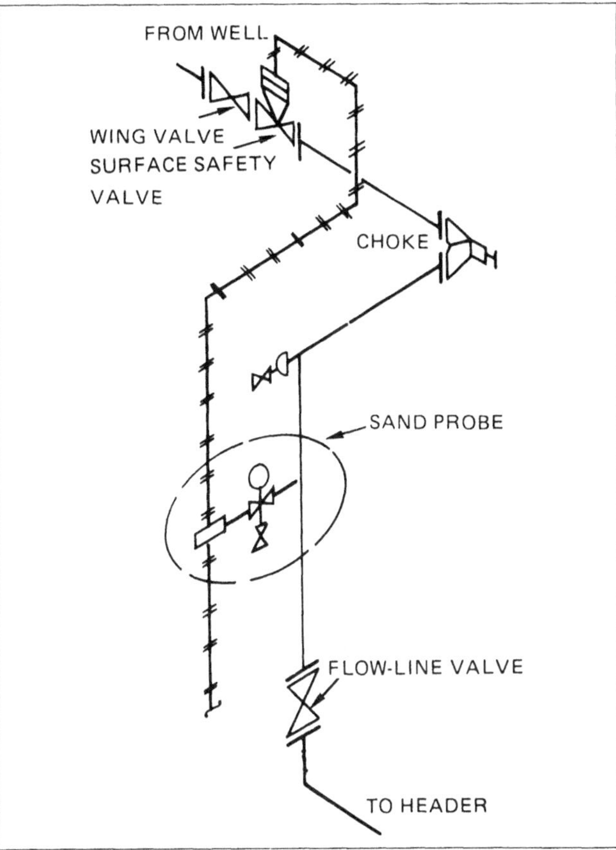

Fig. 2.3—Suggested sand-probe installation on vertical downcourse stream of wellhead choke.

The sand probe has three principal applications. It serves as a safety device, as an aid in optimizing production rates, and as a tool in the selection of workover candidates. It has been instrumental in reducing erosion and has decreased costs associated with equipment failures related to sand erosion. Fire and pollution hazards have also been reduced.

Experience has demonstrated several characteristics about sand erosion and sand probes.

1. The most serious erosion occurs immediately downstream of every change in flow direction on the outside of the turn.
2. Sand returns to the middle of the flow stream about 20 pipe diameters downstream of each turn.
3. Bull-plugged tees are more resistant to sand erosion than elbows are, because the fluid-cushion effect tends to shift the angle of impingement, move the erosion area downstream, and decrease the overall erosion rate.
4. Pipe that changes direction after a long, straight run tends to sustain more severe erosion than pipe with shorter runs of straight flow. The erosion of a given amount of sand-probe wall thickness will correspond to the reduction of a like amount of flowline thickness in the elbows, tees, and valves nearest the sand probe. Consequently, the probes are typically installed on the vertical downcourse downstream of the wellhead chokes (**Fig. 2.3**).

Sand probes have functioned quite well as safety devices. Most of the probes in service offshore have a 0.020-in. wall thickness. The majority of offshore flowlines are 2-in. Schedule 80, with a nominal wall thickness of 0.218 in., or about 10 times that of a sand probe. A minimum wall thickness of 0.150 in. is required to contain the maximum system pressure of 1,200 psi safely. Thus, a new 2-in. Schedule 80 flowline has a safety excess of ≈0.070 in.

Because sand-probe and flowline erosion usually occur at about the same rate, flowline erosion can be considered critical when three sand probes have failed. As a safety precaution, therefore, the practice has been to inspect flowlines ultrasonically or radiographically after every third sand-probe cutout.

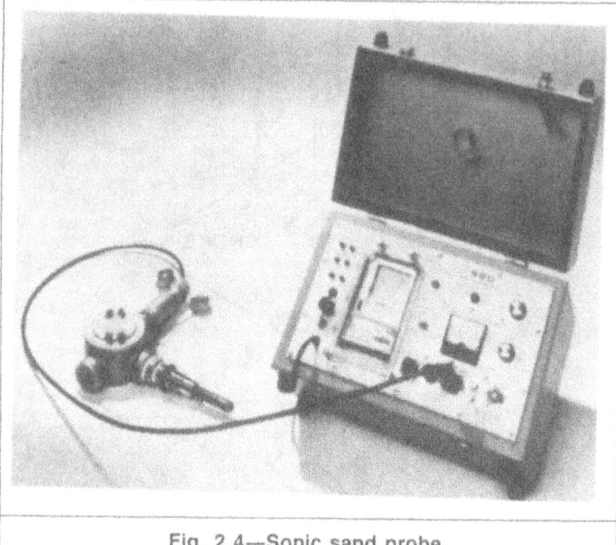

Fig. 2.4—Sonic sand probe.

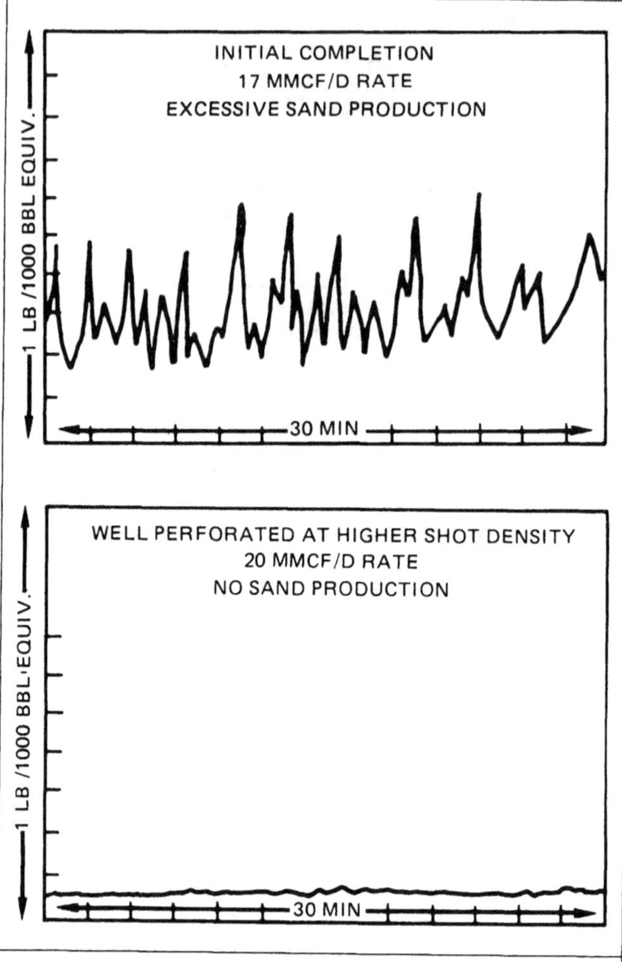

Fig. 2.5—Response of sonic probe before and after perforation of sand-producing well to higher shot density.[10]

Production can be optimized when a sand probe is used. If a well yields sand at a high production rate after producing sand-free at a lower rate, the maximum sand-free production rate can be achieved by reducing the high rate in 20% increments until probes no longer fail. Production is then increased gradually to the maximum sand-free rate.

Sand probes can be used to great advantage in the early selection of workover candidates, thereby reducing workover costs. The need for a workover to control sand production usually becomes evident only when a well stops flowing because it is sanded up. Sand must then be washed out of the well before any remedial work can begin. A sand probe gives advance warning of a potential sandout, thus providing the opportunity for precautionary, less-expensive sand-control treatments.

Sonic Probes. Probes that detect the impact of sand as an acoustical signal are available. These are sometimes called sand detectors.[9] The principal advantage is that the probe provides an instantaneous indication of sand production. Its disadvantages include the expense involved and the fact that the signal currently is not related to the erosiveness of the sand being produced. Its use is restricted primarily to high-rate wells. **Fig. 2.4** is a photograph of the sonic sand probe and instrument package.

The probe is mounted in a surface flowline. Acoustical "pinging" of impinging sand is converted in the probe to an electrical signal proportional to the energy created by the impingement. The signal can be calibrated to determine the concentration of solids in terms of pounds per day as a function of fluid velocity. Both the mass concentration of solids in the flow stream and the rate of sand production are provided. If fluid or gas production rates are known, the rate of sand production and solids concentration can be determined within a factor of two.[10]

Sand concentration as low as 10 lbm/1,000 bbl at flow velocities as low as 3 ft/sec, or the gas equivalent, are said to be detectable by the sonic probe. The accuracy and sensitivity of the equipment are reduced if solids are very fine (i.e., silts) and if the flow system is liquid with severe gas slugging.

A device similar to the sand detector has also been developed for use exclusively with gas production.[11] Like the sonic sand detector, this device detects impingement of sand particles by a piezoelectric crystal but counts each particle impact to determine the sand cut. As a result, the detector is best suited for detecting sand in a dry-gas stream because the impingement of water droplets is recorded as sand particles. Likewise, it is not suited for liquid service where multiphase flow exists.

Sand detectors can be used to determine a well's maximum sand-free production rate and sand production rates at increased flow rates. The effectiveness of various completion, stimulation, and production practices may be assessed and maximum production maintained with these devices. **Fig. 2.5** illustrates the sonic sand detector's response before and after a sand-producing well was perforated to a higher shot density, verifying the effectiveness of the remedy. However, the effect of adding perforations must be normalized to determine whether the actual sand production has decreased. Continuous detection and monitoring in surface flowlines permit corrective action to be taken before excessive erosion damage has occurred, either downhole or at the surface. Also, remedial steps may be taken before there is significant reservoir disruption from sand production.

Painted Pipe. The painted-pipe probe is simply a length of pipe that is coated with an abradable substance.* The coating may be paint or another material. When the probe is placed opposite sand-producing orifices and production is initiated, a portion of the coating is removed by the impingement of sand entrained in the produced fluid, thereby marking the location of sand production. The degree of sand entrainment can be approximated by the amount of coating eroded from the pipe. The tool, which can be run on a wireline, can be used to locate sand-producing strata, perforations, and holes in damaged or corroded screens and casing.

A painted tailpipe is sometimes run below a backsurge tool to indicate perforation effectiveness; i.e., the paint erodes opposite any open perforations. When too few perforations are shown to be open, the interval can be reperforated or the backsurge tool rerun.

Completion Practices

If a well is to be completed in an unconsolidated formation without a sand-control treatment, several completion practices should be followed to minimize the possibility of formation failure and subsequent loss of production. In general, these practices are intended

*Personal communication with G.P. Maly.

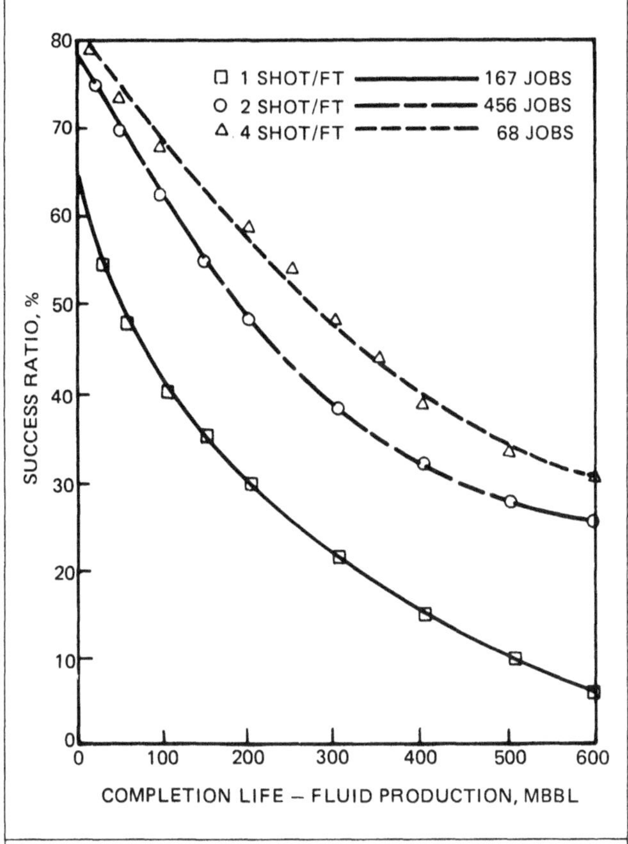

Fig. 2.6—Effect of perforation density on successful production life.

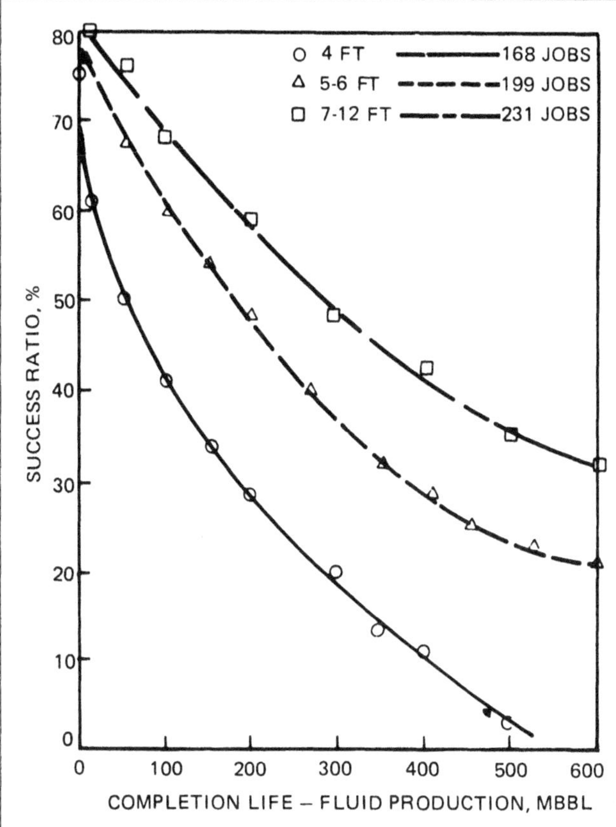

Fig. 2.7—Effect of completion length on successful production life.

to reduce the stresses caused per unit of production by enhancing the ability of the formation to produce fluid rather than sand. They include the use of clean completion fluids, high perforation densities, perforation of long intervals, and perforation of clean sands. The quality of the formation sand (i.e., high or low permeability) also influences sand production.

Completion Fluids. The oil industry has long recognized the need for clean workover and completion fluids. Fluids that are free of solids and compatible with formation material are essential to successful sand control.[12] In addition, every effort should be made to ensure that all perforations are open for production and that the sandface is clean so that the velocity of fluid withdrawal is reduced. Partially plugged perforations and/or formation damage caused by incompatible or solids-laden wellbore fluids can mean excessive pressure drawdowns and stress levels to produce desired fluid rates.

Perforation Density. A study[13] of the effect of perforation density on unconsolidated formation failure revealed that sand problems in untreated intervals could be minimized by increasing perforation density. **Fig. 2.6** shows the results of 691 untreated completions in three offshore Louisiana fields. Average cumulative production, before the sanding of intervals perforated at 4 shots/ft, exceeds 285,000 bbl of fluid. This represents a seven-fold improvement in production life over intervals perforated with only 1 shot/ft. Although 2 shots/ft were far more successful than 1 shot/ft, the average production life is only 66% that of a 4-shots/ft completion.

Wells not only should have high shot densities but also should be perforated with clean fluids with a pressure differential into the wellbore. Plugging of the perforations with gun debris and other solids will thereby be minimized. It is also important to have large-diameter (at least 0.75-in.) perforations in case the well later requires gravel packing.

Calculation of Success. The success-vs.-production curves represent the lifetime of a completion independent of the different production rates prevailing from well to well. Production life is expressed in barrels of produced fluid. Gas production was converted to equivalent liquid production through an empirical factor, 8,000 scf=1 bbl. Success is explicitly sand-control success. Any given job is a failure only if the well sands up or gives some other indication of sand-control failure.

Success at any volume of posttreatment production is represented as the fraction of treated wells producing sand-free. The denominator in the fraction includes only treated wells that could have reached that particular production volume. The total number of wells treated is reduced by the number of previous sand-control failures and by the number of wells that have not had the opportunity to reach a particular production volume. In general, success curves are calculated with

$$S_B=\frac{n_B}{T-\sum_{i-1}^{B}\frac{n_{i-1}^{i}}{S_{i-1}}},$$

where

T = total jobs,
S_B = fractional success at B barrels of production,
n_B = number of jobs that have produced at least B barrels,
i = production increments,
n_{i-1}^{i} = number of jobs that have produced at least $i-1$ but have not yet had the opportunity to produce i increments of production, and
S_{i-1} = fractional success at $i-1$ increments of production.

Interval Length. The frequency of sand problems in wells completed without sand-control measures decreases significantly with increasing length of exposed interval.[13] Results from other offshore completions shown in **Fig. 2.7** support this conclusion. Before sand problems arose, the average production life in completion intervals only 4 ft long had been <60,000 bbl of fluid. Completion intervals of 5 to 6 ft maintained an average life of 180,000

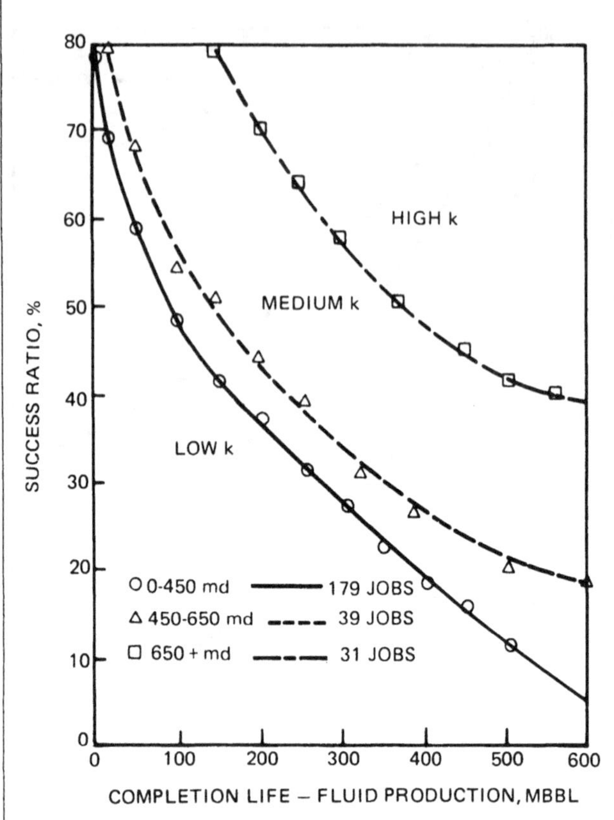

Fig. 2.8—Effect of permeability-thickness product on successful production life.

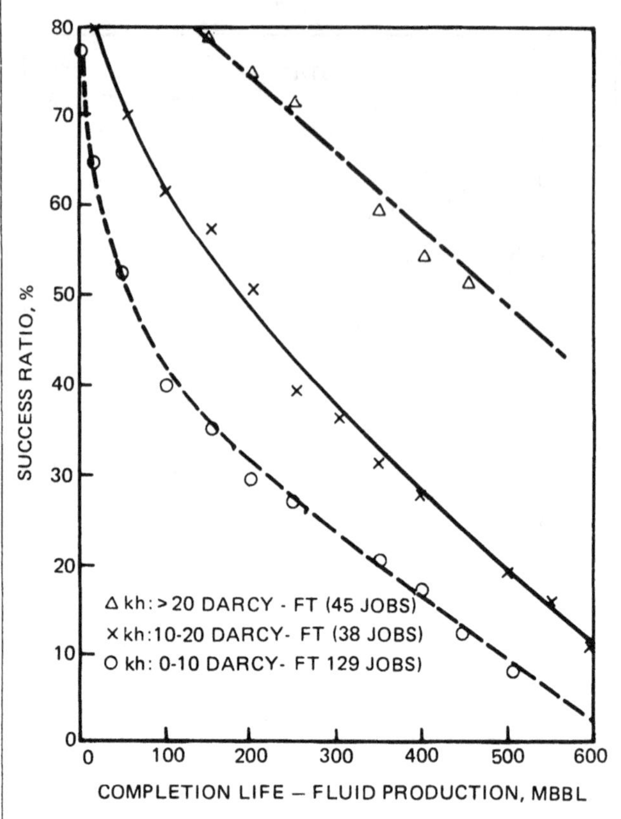

Fig. 2.9—Effect of reservoir permeability, *k*, on successful production life.

bbl, a three-fold increase. Interval lengths of 7 to 12 ft produced an average of five times the fluid of 4-ft intervals before sand problems occurred. Data for interval lengths > 12 ft exhibit very little improved performance compared with the 7- to 12-ft group.

Sand Quality. Sand problems are more severe in dirty, fine-grained formations than in relatively clean, well-developed sands. The data verified that high-permeability formations (usually cleaner and larger-grained sand) were produced more successfully without sand-control techniques than low-permeability zones (usually smaller sand grains with streaks of shale). **Fig. 2.8** shows the effect of reservoir permeability on successful production life, and **Fig. 2.9** shows the effects of the permeability-thickness product.[13]

In addition to reservoir permeability, the cleanliness of the zone perforated within a reservoir influences completion life.[13] Examination of the spontaneous-potential or gamma-ray curve development on 117 completion intervals revealed that zones with good development created fewer sand problems than zones with poor and erratic development. Perforation of poorly developed sections within a good, clean formation should be avoided unless needed for reservoir drainage. If a "ratty" section is to be included in the perforation interval, sand-control measures after initial completion should be considered.

The success of producing the more permeable segments reflects the effect of limiting wellbore stresses by reducing drawdown. High perforation densities and long intervals create the same effect. Also helpful is the fact that segments containing a high percentage of nonsilica particles (indicated by a poorly developed spontaneous-potential curve) generally exhibit low permeability and poor natural cohesion. These intervals should be avoided if possible.

References

1. Morita, N. *et al.*: "Parametric Study of Sand-Production Prediction: Analytical Approach," *SPEPE* (Feb. 1989) 25-33; *Trans.*, AIME, **287.**
2. Morita, N. *et al.*: "Realistic Sand-Production Prediction: Numerical Approach," *SPEPE* (Feb. 1989) 15-24; *Trans.*, AIME, **287.**
3. Halleck, P.M. and Damasena, E.: "Sand Production Tests Under Simulated Downhole Pressure Conditions Using Shaped-Charge Perforated Well Core," paper SPE 19748 presented at the 1989 SPE Annual Technical Conference and Exhibition, San Antonio, Oct. 8-11.
4. Saucier, R.J.: "Considerations in Gravel Pack Design," *JPT* (Feb. 1974) 205-12; *Trans.*, AIME, **257.**
5. Penberthy, W.L. Jr.: "Gravel Placement Through Perforations and Perforation Cleaning for Gravel Packing," *JPT* (Feb. 1988) 229-36; *Trans.*, AIME, **285.**
6. Tixier, M.P., Loveless, G.W., and Anderson, R.A.: "Estimation of Formation Strength From the Mechanical-Properties Log," *JPT* (March 1975) 283-93.
7. Stein, N., Odeh, A.S., and Jones, L.G.: "Estimating Maximum Sand-Free Production Rates From Friable Sands for Different Well Completion Geometries," *JPT* (Oct. 1974) 1156-58; *Trans.*, AIME, **257.**
8. Swan, R.D. and Reimer, C.M.: "The Development and Use of Sand Probes," paper SPE 4555 presented at the 1973 SPE Annual Meeting, Las Vegas, Sept. 30-Oct. 3.
9. Mullins, L.D., Baldwin, W.F., and Berry, P.M.: "Surface Flowline Sand Detection," paper SPE 5152 presented at the 1974 SPE Annual Meeting, Houston, Oct. 6-9.
10. Foster, C.R. and Linville, T.W.: "A Method of Monitoring Sand Production in a Flowing Well Stream," paper SPE 8214 presented at the 1979 SPE Annual Technical Conference and Exhibition, Las Vegas, Sept. 23-26.
11. Stuivenwold, P.A. and Mast, H.: "Sand Detector: Novel Instrumentation for Managing Sand-Problem-Prone Fields," paper SPE 9368 presented at the 1980 SPE Annual Technical Conference and Exhibition, Sept. 21-24.
12. Suman, G.O. Jr., Ellis, R.C., and Snyder, R.E.: *World Oil Sand Control Handbook*, second edition (1975) 11.
13. Schroeder, R.H. and Tucker, M.J. III: "Evaluation of Completion Practices for Improved Sand Control Success and Longevity," paper SPE 5027 presented at the 1974 SPE Annual Meeting, Houston, Oct. 6-9.

SI Metric Conversion Factors

bbl	× 1.589 873	E−01	=	m^3
ft	× 3.048*	E−01	=	m^3
ft^3	× 2.831 685	E−02	=	m^3
in.	× 2.54*	E+00	=	cm
lbm	× 4.535 924	E−01	=	kg
psi	× 6.894 757	E+00	=	kPa

*Conversion factor is exact.

Chapter 3
Gravel-Pack Productivity

Introduction

Controlling sand in wells originated with the water-well industry because early water-well completions in shallow formations commonly produced sand. Gravel packing was introduced in the water-well industry in the early 1900's, but the technique was not used in oil wells until the early 1930's. It involves placing accurately sized coarse-grained material (gravel) against the formation sand to prevent the production of the finer-grained material while fluids are produced. A screen/slotted liner is located concentrically inside the layer of gravel to prevent gravel entry into the well. In the early completions, the mechanical devices included torch-cut slots, perforated casing, a louver-type screen, and machine-slotted pipe. More recently, several varieties of wire-wrapped screen have been used for this purpose. (See Chap. 5 for a detailed discussion of screens and slotted liners.) The technology used in gravel packing oil and gas wells was borrowed from the water-well industry. Unlike most oil and gas wells, water wells usually were completed by developing a natural gravel pack, which involved alternately surging and producing the formation. The finer particles were ultimately produced from the well, which left the uniformly graded sand or gravel with higher porosity and permeability surrounding the screen or slotted liner.

Effective development of a water well by this method might take several weeks, and the effectiveness depends on the sand characteristics, well-screen design, and the driller's skill. **Fig. 3.1** is a schematic of a natural gravel pack. Water wells that were not developed to form a natural gravel pack but needed sand control usually were gravel-packed.

Because the gravel-packing technology was borrowed, early oil and gas completions were similar to those of water wells and at first included running only a slotted liner to prevent sand production. Openhole gravel packing was the most commonly used alternative if slotted liners proved ineffective. Oil- and gas-well completions opened a new dimension in gravel packing. Perforated-casing completions became popular with the advent of gun and jet perforating. In these cases, screens were run inside a well's casing and gravel was packed around them.

Gravel-Packed Completions

Gravel packing consists of installing a downhole filter to exclude formation sand. **Fig. 3.2** shows the two basic gravel-pack geometries currently in use, the openhole and cased-hole techniques.

Openhole Gravel Packs. In openhole gravel packs, there is no casing between the gravel pack and the formation sand (see Fig. 3.2). This completion has the highest productivity of all gravel-packed completions and is surpassed only by openhole completions.

The primary disadvantage of the openhole gravel pack is the difficult task of isolating extraneous fluids, such as gas or water, at the sandface. A second limitation is that not all sand formations are physically structured to accommodate this type of completion, causing hole instability. In these cases, the sand formations are usually cased off before completion.

The openhole gravel pack is the preferred mechanical sand control technique for high-productivity wells. It is most effective in long-life completions where hole stability or the production of water and/or gas will not present a problem.

Cased-Hole Gravel Packs. The cased-hole gravel pack consists of a screen or slotted liner that is gravel packed inside perforated casing (see Fig. 3.2). This completion is the most widely used gravel pack in oil and gas wells because, upon initial completion, operators are not always certain whether wells will produce formation sand. Consequently, the casing is perforated and the wells are produced. If no sand appears, a gravel pack is not necessary. A cased-hole gravel pack can be performed later, however, if sand production is not manageable. Furthermore, this completion is better adapted to exclude water and/or gas if the need arises. Drilling and workover considerations often dictate that a cased completion is more desirable. For these reasons, cased-hole gravel packs are more common than openhole gravel packs.

The primary disadvantages of the cased-hole gravel pack are that it is not as well suited for high flow rates as the openhole gravel pack and that it is much more difficult to perform correctly. These limitations arise from fluid flow through sand-filled perforations. The cross-sectional area through the perforations represents only a few percent of the area exposed in openhole completions, even at extremely high perforation densities (such as 24 shots/ft). Therefore, it is important that all perforations and the regions outside them contain gravel of the highest permeability that will control formation sand effectively. Gravel packing outside the perforations is commonly referred to as prepacking but should not be confused with prepacked screens (discussed in Chap. 5). Prepacking is important for achieving high-productivity, long-life completions.

Table 3.1[1] compares open- and cased-hole productivities. Note that, for this example, the openhole gravel pack has a much higher productivity than the cased-hole gravel packs. Also, the cased-hole gravel pack, with gravel placed through and outside the perforations, has a higher productivity than the case without prepack. For

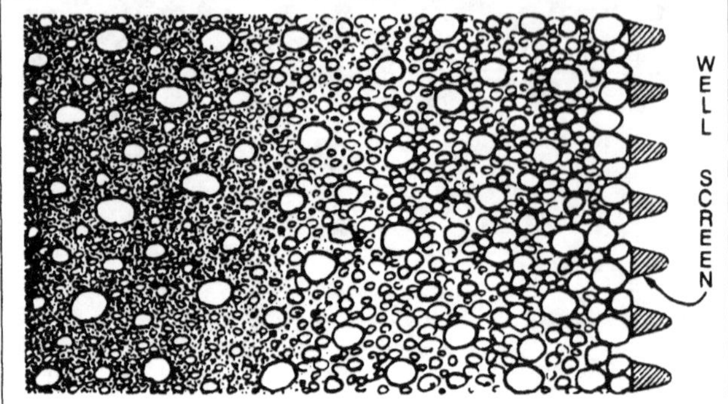

Fig. 3.1—Schematic of a natural gravel pack after developing (courtesy Johnson Div., UOP Inc.)

TABLE 3.1—GRAVEL-PACK PRODUCTIVITY [1]

	PI (BOPD/psi)	
	Well LL-5*	Well LL-3*
Openhole gravel pack	48.4 (14)	6.4 (13)
Gun-perforated casing	36.6 (20)	5.2 (14)
Cased-hole gravel pack (with prepack)	12.9 (19)	3.2 (12)
Cased-hole gravel pack (without prepack)	4.0 (14)	1.7 (3)

*Miocene reservoirs, Venezuela. ( ) = number of wells.

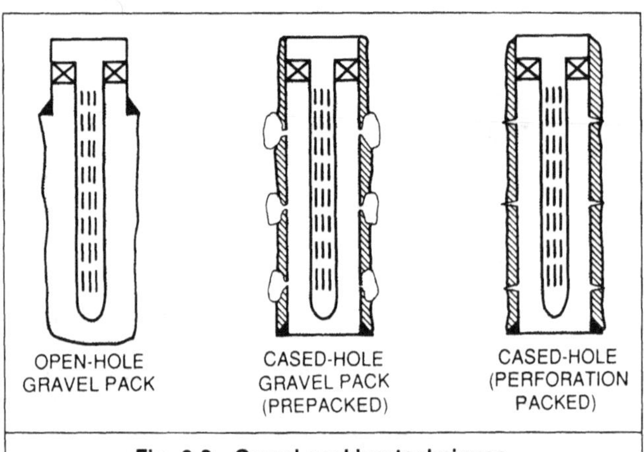

Fig. 3.2—Gravel-packing techniques.

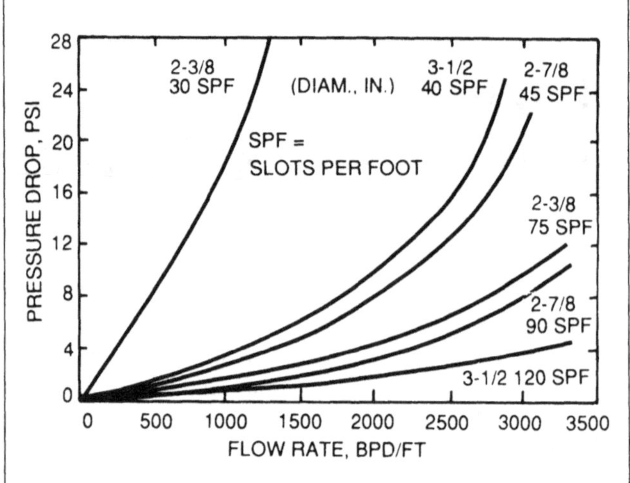

Fig. 3.4—Flow capacity of slotted liners, 0.020-in. slots, with 20/40-U.S.-mesh gravel.

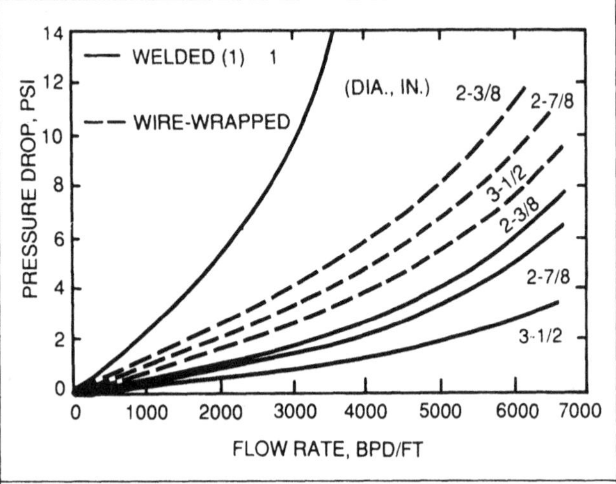

Fig. 3.3—Flow capacity of 12-gauge screens with 20/40-U.S.-mesh gravel.

this example, the cased-hole gravel packs were perforated from 4 to 6 shots/ft and do not reflect the productivities possible with current perforating technology. Productivities would have been higher had high-density, large-diameter perforating practices been used.

As a result of problems associated with gravel-pack productivity, a whole new dimension in gravel packing arose with the cased-hole gravel-pack completion because substantially different techniques and procedures are required. Sand control is not difficult in cased-hole gravel packs, but in many instances the wells are rate-limited because improper techniques and procedures were used. By contrast, openhole gravel packs are not limited by flow through gravel-filled perforations.

The productivity limitations encountered in cased-hole gravel packs prompted numerous investigators to begin work to define and solve related problems. Early work on the proper gravel size required for effective sand control indicated that gravel/sand ratios from 5 to 10 should be used. Design points anywhere from 10 to 70% cumulative were recommended.[2-13]

Most of this research was conducted in linear flow packs operated under single-phase flow conditions. Additional work[8-10,13] confirmed that the main restriction to flow was sand-filled perforations. For maximum productivity, plugging the perforation with formation sand must be avoided. The solution has been to pressure-pack the perforations with gravel of the highest permeability capable of preventing the production of formation sand.

New emphasis has been placed on gravel packing, particularly when cased-hole packs are required. This interest has evolved because of the new high-rate fields under development that will require sand control. Examples are fields in the North Sea, the Middle East, and Southeast Asia. Some of the wells in these areas are capable of producing at rates exceeding 200 BFPD/ft. Research on cased-hole gravel-pack performance has also accelerated because of higher water and gas cuts associated with produced oil in older fields, a circumstance tending to have adverse effects on productivity and sand production.

Completion Productivity

Studies have been conducted to determine the magnitude of productivity losses associated with screens and slotted liners with gravel packed around them. Test results (**Figs. 3.3 and 3.4**) show that the flow rates were extremely high, but pressure losses were quite low. Wire-wrapped screens had higher flow capacities than slotted liners; however, at realistic field rates, the pressure drop through the screens or liners was small. Note also that the flow capacity of slotted liners is proportional to the number of slots per foot, and not to the liner diameter as with screens.* See Chap. 5 for a discussion of screens and slotted liners.

*Unpublished Exxon Production Research Co. report, Houston (1978).

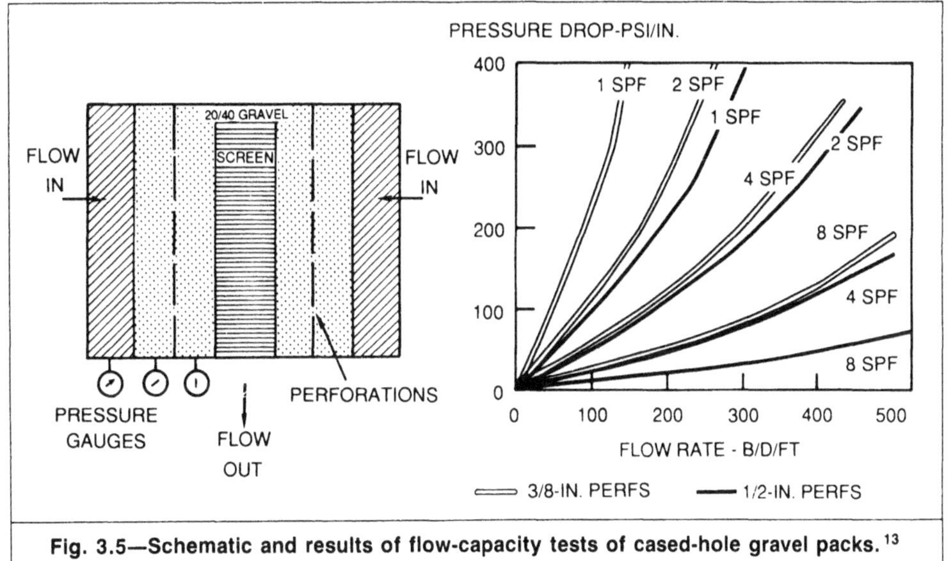

Fig. 3.5—Schematic and results of flow-capacity tests of cased-hole gravel packs.[13]

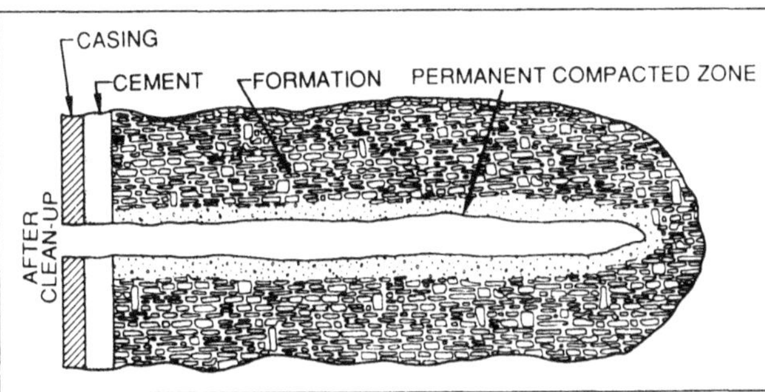

Fig. 3.6—Schematic of a perforation after cleanup (courtesy Schlumberger).

The implication of these tests is that the screen/slotted liner and gravel pack offer no significant restriction to well productivity, unless they become plugged. Slotted liners seem to become plugged more easily than wire-wrapped screens.

Fig. 3.5[13] shows results obtained from simulated cased-hole gravel packs. These results demonstrate that well productivity may be significantly reduced in this type of completion but that increasing the number and size of perforations can improve productivity significantly. Comparison of the rates shown in Figs. 3.3 and 3.4, which reveal only small pressure drops, with these results shows that a perforated gravel pack is productivity-limited by virtue of the flow through the sand- or gravel-filled perforations. This suggests that perforating practices are very important in cased-hole gravel-pack completions.

Perforating

Most cased-hole gravel-packed wells are perforated with jet perforators. Jet perforating consists of firing a shaped charge to generate an impingement pressure of about 5 million psi on the inner-casing wall and formation. Because this pressure is much higher than the yield strength of either the casing wall or the formation, a hole and tunnel are created with dimensions related to the charge design and the amount of explosive. Shaped charges can be designed to create either long, small-diameter perforation tunnels or short, large-diameter tunnels. The high pressures generated by the jet actually crush and compact the reservoir rock in a zone adjacent to the perforation tunnel, as **Fig. 3.6** illustrates.

Most investigations[14-16] indicate that, with conventional perforating practices, a perforation density of about 4 shots/ft with a penetration of about 6 to 8 in. results in a productivity equivalent to that of an openhole completion (**Fig. 3.7**). Perforating a well to be gravel-packed yields different results because, after gravel packing, the perforating tunnel is filled with gravel. At a gravel porosity of about 35%, roughly two-thirds of the cross-sectional

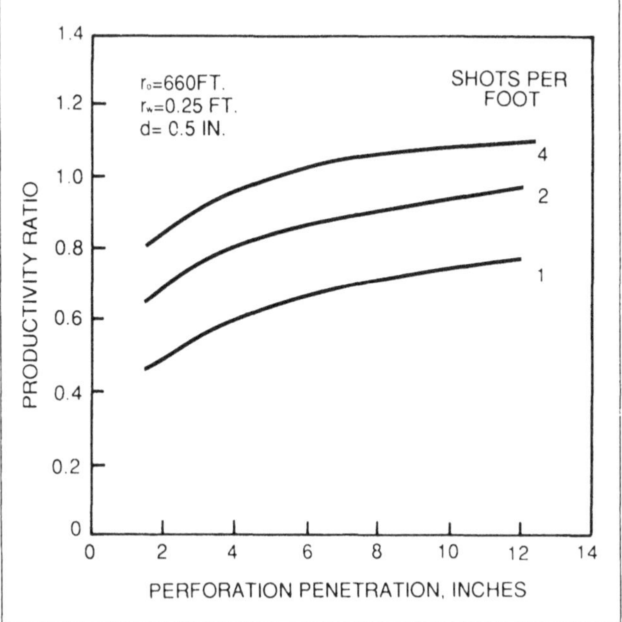

Fig. 3.7—Effect of penetration and shot density on productivity (after Harris[15]).

area open to flow contains gravel. Comparison of **Figs. 3.8** and 3.6 shows the impact of gravel-filled perforations. The implications here are that higher perforation densities are necessary for gravel-packed completions than for conventional completions. Higher perforation densities are beneficial because linear flow through the gravel-filled perforation can cause high pressure losses and reduced well productivity.

Large-diameter perforations are also desirable to expose more inflow area to the well. For gravel packing, large-diameter perforations are more important than perforation length, provided that the perforations effectively communicate with the formation. Perforation diameters from 0.75 in. to >1 in. may be required for high well productivity. In evaluations of perforation charges, test results should conform to API *RP 43, Standard Procedure for Evaluation of Well Perforators.*

To arrive at the optimum completion configuration,[17] designs of a perforating program for a gravel pack must consider the fluid flow from the reservoir, through the perforations, and from the well. Some operators have used a systems approach to gravel-pack completion design called nodal[18] analysis that provides a reasonably sound basis for determining the tubing size and perforation requirement. For particular fields and operating conditions, however, this

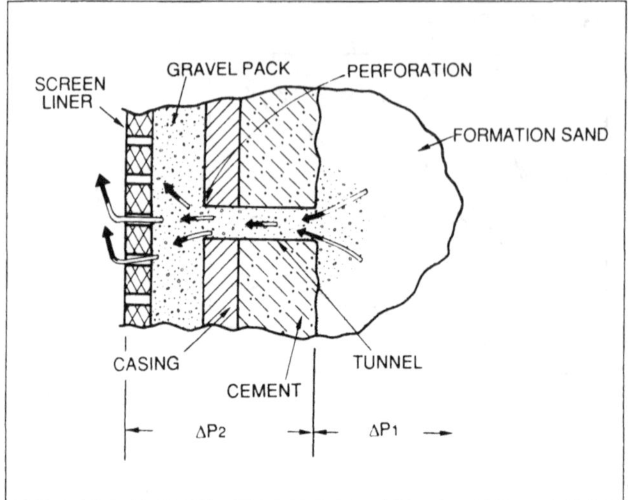

Fig. 3.8—Schematic of an ideally gravel-packed completion (after Bell[14]).

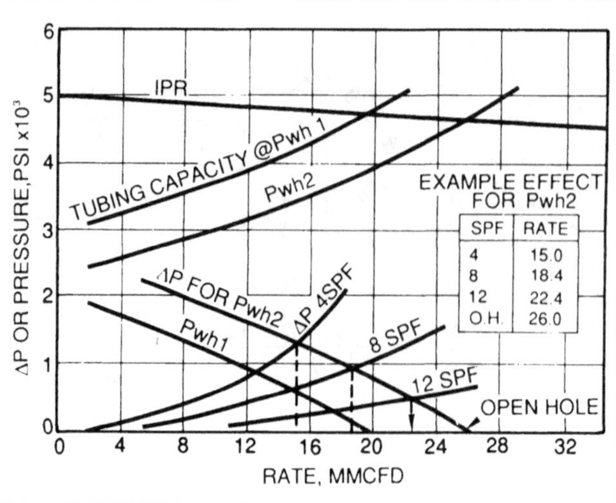

Fig. 3.9—Systems response curves to gravel-pack perforation design (after Crouch and Pack[17]).

approach may give optimistic production forecasts if the perforations are damaged or yield small-diameter holes. This design approach is illustrated in **Fig. 3.9**. The logic used in this approach is that the difference between the well's inflow and outflow determines the amount of pressure loss that can be taken across the perforations. The inflow [PI or inflow-performance-relationship (IPR)] curves are usually calculated first on the basis of reservoir test data. The tubing capacity is then determined as a function of wellhead pressure for a given tubing size. By subtracting the tubing-capacity curve from that of the PI or IPR curve and replotting it as the pressure drop across the completion, the allowable pressure drop through the perforations is determined. The pressure drop through the perforations can then be calculated with either Eq. 3.2 or 3.3 and plotted as shown in Fig. 3.9 to arrive at the proper perforation size and density configuration for that particular flow rate. The intersection of the pressure drops for the various perforation sizes and densities with the replotted curves is the highest flow rate for that perforation size and shot density. However, should the pressure drop across a perforation exceed a value known to create problems in the field, additional or larger perforations should be added. For example, a well is known to be capable of flowing 20 Mscf/D at a wellhead pressure of p_{wh2}. What perforation diameter and shot density are required? For 0.7-in.-diameter perforations, Fig. 3.9 indicates that a minimum of 10 shots/ft is necessary to accommodate this rate.

Because the perforation tunnels can be a source of high pressure drop, particularly if formation sand enters them, it is extremely important that the proper size and number of perforations are selected. Calculations (see Eq. 3.2 or 3.3) usually reflect ideal or optimistic projections of the pressure drop across perforations because the actual gravel permeability may be lower than that for a given mesh size owing to the generation of fines during handling. Also, the presence of gas or water saturation in the gravel, the invasion of formation fines, or the gravel being oil-wet tends to reduce its permeability. The perforation size could also be smaller than that specified. Finally, some of the perforations may not be contributing to flow because they are plugged or otherwise damaged. Hence, this procedure provides a way to determine the tubing size and the minimum perforating requirement for a gravel-packed completion but should be coupled with field experience.

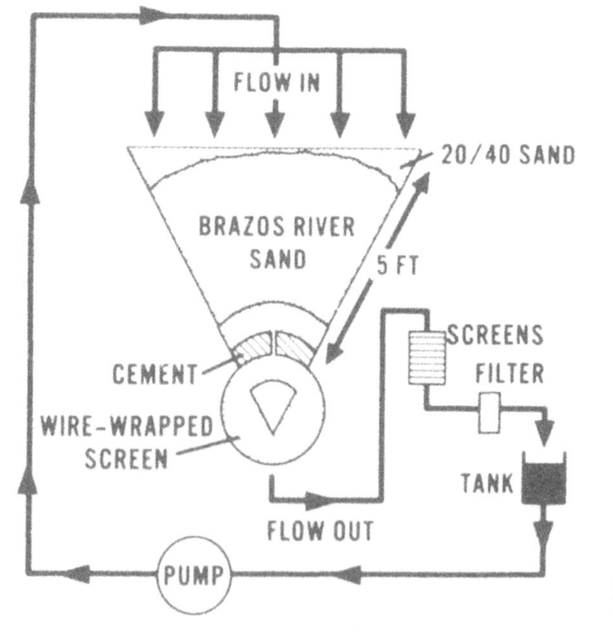

Fig. 3.10—Model and schematic of full-scale gravel-pack productivity tests.

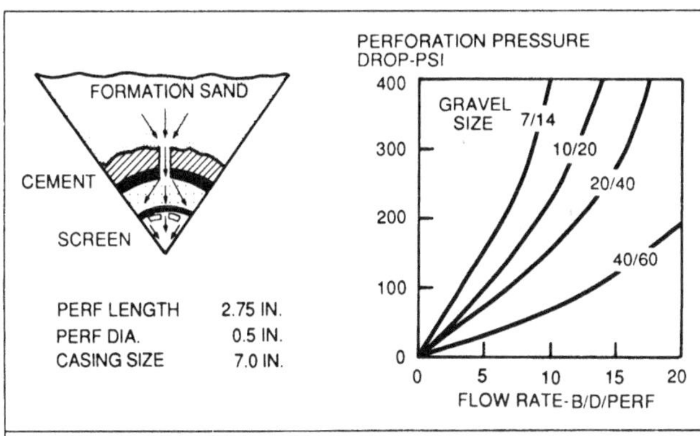

Fig. 3.11—Results of cased-hole gravel-pack tests (no prepack—perforation filled with gravel).[13]

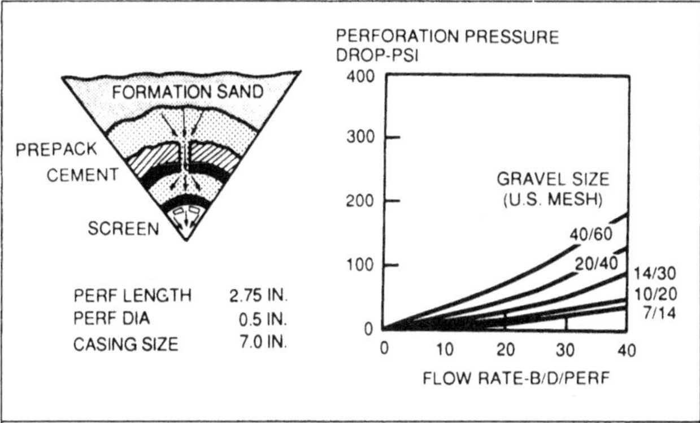

Fig. 3.12—Results of cased-hole gravel-pack tests (2-in. prepack).

In many cases, the pressure drop through the tubing and the reservoir productivity are already known. The perforating program for the particular well is merely the one that provides the optimum perforation pressure drop for the projected flow rate from a cost and operational standpoint.

Equations useful in calculating the linear pressure drop through a single perforation are derived from the Forchheimer equation (Eq. 3.1 shows the general equation), which is usually used to describe non-Darcy (turbulent) flow[19]:

$$\Delta p_L = (\mu v L/k) + \beta \rho v^2 L, \quad \text{(3.1)}$$

where

Δp_L = pressure drop, psi,
μ = viscosity, cp,
v = Darcy superficial velocity, cm/sec,
L = length of perforation tunnel, ft,
k = permeability, darcies,
β = turbulence factor, atm-sec²/g, and
ρ = density, cm³.

The first term in Eq. 3.1 accounts for the pressure drop caused by laminar flow; the second term describes the pressure drop caused by turbulence.

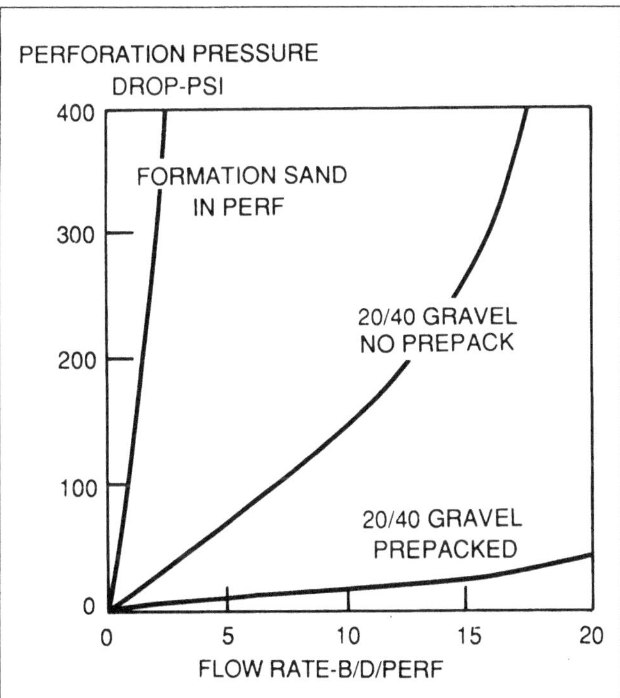

Fig. 3.13—Productivity with prepacking, without prepacking, and with formation sand in the perforations.

Eq. 3.1 can be expressed in engineering units for liquid flow.[9]

$$\Delta p_L = 0.888(L\mu q/kA) + 9.1\times 10^{-13}\beta L\rho(q/A)^2, \quad \text{(3.2)}$$

where

k = permeability of tunnel fill material, darcies,
q = flow rate per perforation, B/D,
A = cross-sectional flow area of perforation tunnel, ft²,
ρ = fluid density, lbm/ft³, and
β = beta factor (inertia coefficient for sandstone) $= 10^{(6.5-9.5 \log k)}$, 1/ft.

For gas flow, Eq. 3.1 becomes, in engineering units,[20]

$$p_2 = 14.7\left\{\left(\frac{p_1}{14.7}\right)^2 - \left[\left(2.57\times 10^{-12}\times\frac{q_g\gamma\beta}{d^2\mu} + \frac{1}{k}\right)\right.\right.$$

$$\left.\left.\times 0.33\times\frac{zT\mu Lq_g\gamma}{d^2}\right]\Big/\left(\gamma 28.964\right)\right\}^{1/2}, \quad \text{(3.3)}$$

where

q_g = gas rate per perforation, scf/D,
γ = gas specific gravity (air = 1.0),
d = perforation diameter, in.,
T = temperature, K, and
z = gas compressibility factor.

Eqs. 3.1 through 3.3. can be used to determine the pressure drop across a perforation containing a gravel of known permeability as a consequence of viscous and turbulent flow. β accounts for the effect of turbulent flow in each equation. The pressure drop increases significantly compared with the viscous pressure drop once turbulent flow is encountered. The design of a gravel pack is also based

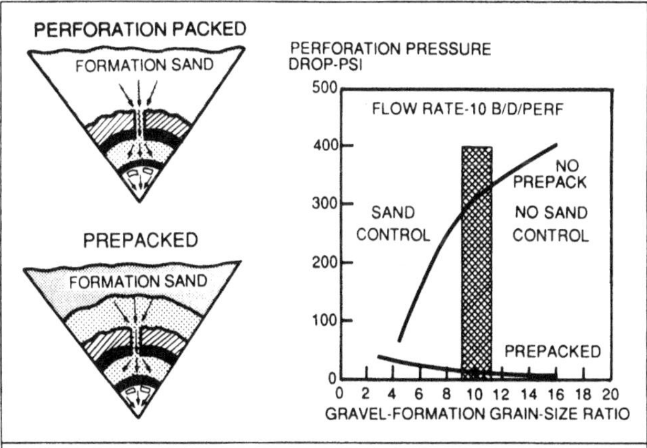

Fig. 3.14—Importance of proper gravel size and prepacking in terms of productivity.

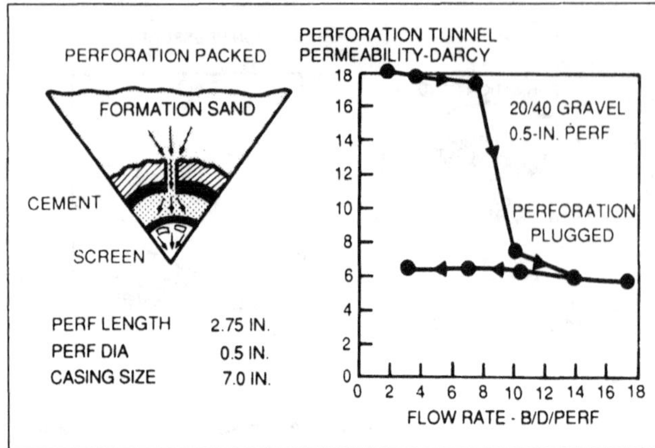

Fig. 3.15—Test results indicating liability of permanent formation damage when prepacking is not used.

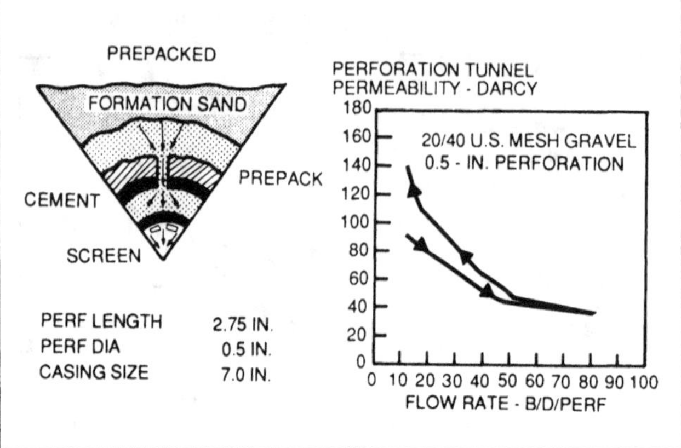

Fig. 3.16—Test results demonstrating importance of prepacking in prevention of perforation damage.

on fluid inflow into the well, so if the pressure drop is too high (or low), the perforation program can be changed to increase (or decrease) the size and number of perforations to meet the design conditions.

Radial Model Gravel-Pack Productivity Studies

The perforation tunnel is probably the most critical region in a cased-hole gravel pack from the standpoint of completion flow capacity. In addition to perforation diameter, the effects of gravel size, gravel/sand geometries, flow rate, and multiphase flow through gravel-filled perforations have been shown to affect well productivity and completion performance significantly. The impact of these factors on productivity and sand control has been studied in the large radial model shown in **Fig. 3.10.**[13] It represented a 70° sector about a well and had a 5-ft radius. Factors that affect gravel-pack productivity could be isolated with this large model and observed in a parametric study more effectively and economically through an analysis of field results.

Uniform Formation Sands. Flow test results with a uniform 5-darcy sand with no prepack but with the perforation filled with gravel (**Fig. 3.11**) revealed a large pressure drop across the perforation. The formation sand had a uniformity coefficient of 1.5 as determined by the ratio of the 40th- and 90th-percentile diameters, d_{40}/d_{90}. The four different gravel sizes used represented gravel/median-sand size ratios of 4.3, 6.9, 9.6, and 14.8. The plugging observed resulted from viscous forces causing the formation sand to migrate into the entrance of the perforation. When a prepack was used (**Fig. 3.12**), however, the pressure drop across the perforation was substantially smaller. The smallest pressure drop was associated with the largest gravel tested (7/14 U.S. mesh).

A comparison between the two types of gravel placement (**Fig. 3.13**)—i.e., with and without prepacking—shows that the smallest pressure drop is associated with the prepacked completions. When no prepack was used, plugging was more severe with larger gravel because the formation sand could migrate farther into the gravel in the perforation. If properly prepacked, on the other hand, formation sand was not produced at concentrations sufficient to cause plugging in the perforation, regardless of gravel size. Finally, the worst condition resulted when the perforations became completely filled with formation sand.

The data also indicated that packing the perforation was not the only requirement for a successful high-rate completion, as **Fig. 3.14** shows. Selecting the proper gravel size was also critical. Again, the large pressure losses were associated with the lack of prepacking, but productivity was improved significantly by prepacking. Fig. 3.14 also shows that, when gravel/median-sand size ratios exceed about 10, sand control becomes less effective as the gravel size increases. Testing indicated that a ratio of about 5 to 6 was optimum for effective sand control with this simulated formation sand. Under these highly controlled conditions, however, small amounts of formation sand were continually produced, regardless of the gravel size used. All data indicated that finite sand production was inherent in gravel packs.

Subsequent testing with no prepack showed that such completions were more rate-sensitive than prepacked completions and that, as a result, permanent damage can result if the threshold rate was exceeded. This liability is demonstrated in **Fig. 3.15.** Plugging or damage occurred when the viscous forces were high enough to move the formation sand into the perforation.

On the other hand, cased-hole prepacked gravel packs are not as severely affected by rate, as **Fig. 3.16** shows. They are affected by turbulent or non-Darcy flow effects. In Fig. 3.16 the apparent perforation-tunnel permeability decreased as the flow rate increased. The apparent loss in permeability was the result of turbulent flow effects. Note that when the flow rate was decreased to the initial level, no permanent damage was evident. The remedy to this apparent loss in permeability is to increase the size and number of perforations.

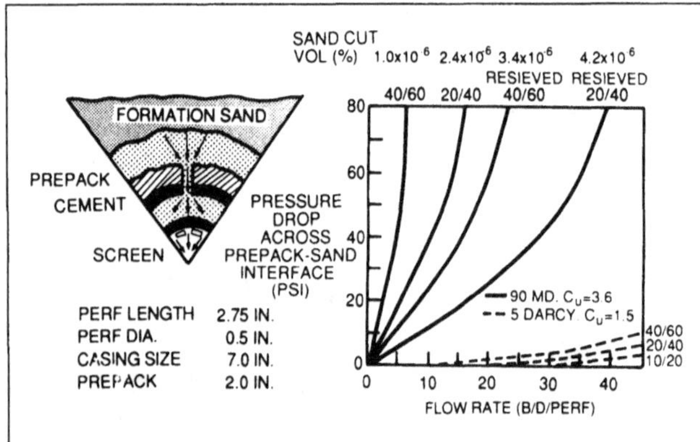

Fig. 3.17—Test results demonstrating importance of gravel uniformity in maintaining optimum productivity.

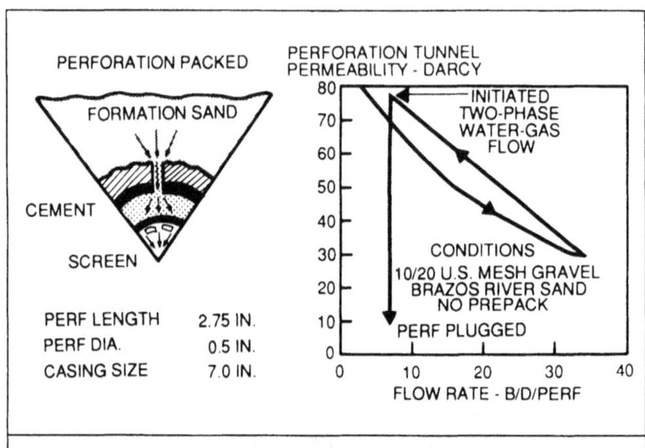

Fig. 3.18—Effect of multiphase flow on productivity.

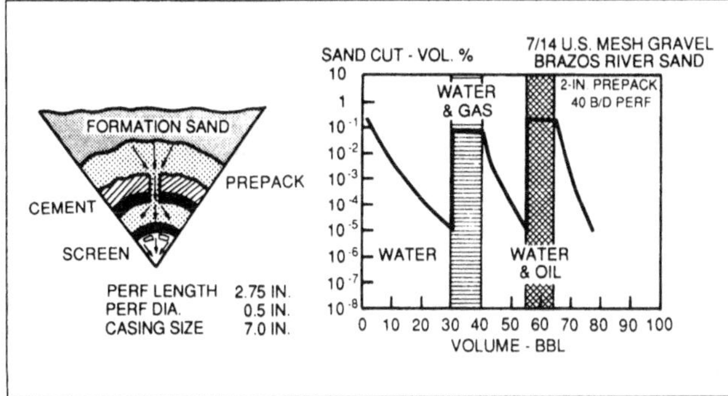

Fig. 3.19—Effect of multiphase flow on formation sand production (cased hole with prepack).

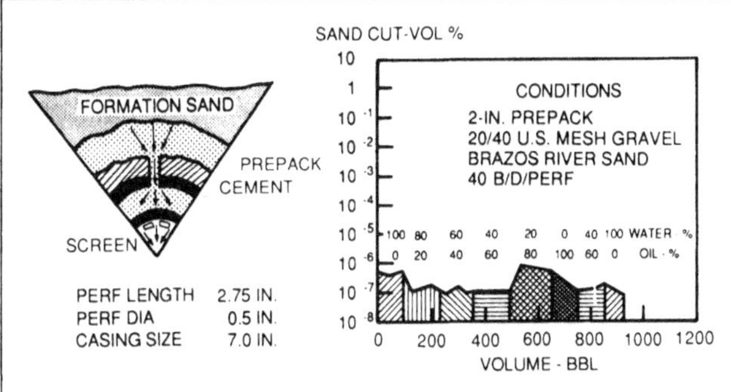

Fig. 3.20—Effect of multiphase flow on formation sand production (cased hole with prepack).

Nonuniform Formation Sands. Tests with a nonuniform sand (uniformity coefficient of 3.6) yielded results significantly different from those with uniform sands. **Fig. 3.17** shows that pressure dropped significantly compared with the uniform sand and that the largest drop occurred at the gravel/sand interface rather than across the perforation. However, the pressure drops across the perforation for the uniform and nonuniform sands were essentially the same.

In these tests, the 40/60- and 20/40-U.S.-mesh sand represented gravel/median-sand size ratios of 3.6 and 6.4. Resieving the gravel decreased the pressure drop across the interface. It also affected the sand cut. All tests with uniform sand involved gravel that was not resieved, yet the pressure drops at the gravel/sand interface were substantially lower. In this case, the sand that was not resieved contained about 10% finer-grained material, which is significantly higher than the 2% fines allowed by API *RP 58, Recommended Practices for Testing Sand Used in Gravel-Pack Operations* (March 31, 1986).

X-ray diffraction analysis indicated that the uniform sand contains 6.9 wt% clay and the nonuniform sand contained 16.5 wt%. The additional pressure drop noted at the gravel/sand interface in Fig. 3.17 resulted because of the increased clay content and the wider particle-size distribution that must bridge when nonuniform sands are gravel packed.

Multiphase-Flow Effects. The effects of multiphase flow on gravel-pack productivity and sand production can be significant. **Fig. 3.18** shows the results of a flow test where 10/20-U.S.-mesh gravel was used to control the uniform sand. This size represents a gravel/median-sand size ratio of about 10. The test results demonstrate that, as long as single-phase water is flowed, there is no permanent damage from perforation plugging. As shown earlier (Fig. 3.15), an apparent permeability loss resulted because of turbulent flow; when the rate was decreased, however, no permanent damage was noted. When two-phase water/gas flow was initiated, permanent plugging of the perforation resulted.

Fig. 3.19 demonstrates prepacked gravel-pack performance when 7/14-U.S.-mesh sand (gravel/sand ratio of 14.8) is used to control formation sand. Note that initial sand production was high but declined rapidly until two-phase flow was initiated, which caused an instantaneous surge in sand production. When single-phase flow conditions were re-established, however, the sand cut decreased, much as it did at the beginning of the test. Fig. 3.19 indicates that it makes little difference whether water/gas or water/oil is flowing; the effect on sand cut is about the same in either case. These examples are intended to demonstrate that multiphase-flow effects can cause and accelerate the production of formation sand. While many different situations in the field may also promote sand production, it is not uncommon for wells to produce sand-free until water production begins. This common event significantly increases the tendency for sand production.

If properly sized gravel is used (gravel/median-sand size ratio of 5 to 6), the effects of multiphase flow are not as severe as those illustrated in **Fig. 3.20**. Although there were variations in sand cut, they were much lower than when larger gravel was used. These test results imply that multiphase-flow effects can cause permanent perforation plugging and influence sand cut if proper gravel sizes are not used.

Gravel-pack experience under numerous conditions has shown that wells commonly decline in productivity, or begin to produce excessive amounts of sand, when water production begins. There are claims that water dissolves cementing material, and that changes in capillary pressure and interfacial tension cause fines migration and sand production. The data presented here do not completely support these claims.

These data and results of other research[21-23] on fines migration indicate that multiphase flow causes localized pressure disturbances because of the numerous interfaces moving through the pore structure and causing excessive fines movement. Results have also shown that, because most fines are water-wet, they do not begin to move until the water phase becomes mobile. These data indicate that, in all probability, a combination of localized disturbances and water-wet fines causes most sand production and loss of productivity in a gravel-packed region.

Summary

The following general conclusions concerning gravel-pack productivity apply during the design and implementation of gravel-packed completions.

1. The screen/slotted liner and gravel pack do not significantly restrict well productivity unless they become plugged with formation sand.
2. For cased-hole gravel packs, maximizing the size and number of perforations increases productivity.
3. Prepacking (sometimes called pressure packing) is essential for high-rate, long-life, cased-hole gravel-packed completions.
4. Turbulence, a major factor in decreasing the productivity in high-rate, cased-hole gravel packs, can be lessened by the use of large-diameter, high-density perforations.
5. Although no single design point completely describes a formation sand, the median, or 50th percentile, point appears to be a compromise (see Chap. 4).
6. A gravel/median-sand size ratio of about 6 yielded the best results for these test conditions. However, ratios from 4 to 8 have also been used in the field successfully for sands with different size distributions and clay and fines contents. Small amounts of formation sand are continually produced through the gravel pack under these conditions.
7. Gravel packs should be designed to allow the production of fine particles. Impedance of this process can result in a loss of productivity.
8. Nonuniform sands are more difficult to gravel pack effectively to maintain high well productivity because of their wide range of grain sizes, which reduces bridging efficiency.
9. The use of accurately sieved gravel may be critical to well productivity.
10. Multiphase flow causes localized pressure disturbances, which can precipitate excessive fines migration, sand production, and perforation plugging.

References

1. Liebach, R.E. and Cirigliano, J.: "Gravel Packing in Venezuela," *Proc.*, Seventh World Pet. Cong., Mexico City (1967) Sec. III, 407-18.
2. Schwartz, D.H.: "Successful Sand Control Design for High-Rate Oil and Water Wells," *JPT* (Sept. 1969) 1193-98.
3. Maly, G.P. and Krueger, R.F.: "Improper Formation Sampling Leads to Improper Selection of Gravel Size," *JPT* (Dec. 1971) 1403-08.
4. Sage, B.H. and Lacy, W.N.: "Effectiveness of Gravel Screens," *Trans.*, AIME (1942) **146,** 89-106.
5. Hill, K.E.: "Factors Affecting the Use of Gravel in Oil Wells," *Drill. and Prod. Prac.*, API (1941) 134-43.
6. Coberly, C.J. and Wagner, E.M.: "Some Considerations in the Selection and Installation of Gravel Packs for Oil Wells," *Pet. Tech.* (Aug. 1938) 1-20.
7. Tausch, G.H. and Corley, C.B. Jr.: "Sand Exclusion in Oil and Gas Wells," *Drill. and Prod. Prac.*, API (1958) 66-82.
8. Williams, B.B., Elliott, L.S., and Weaver, R.H.: "Productivity of Inside Casing Gravel-Pack Completions," *JPT* (April 1972) 419-25.
9. Saucier, R.J.: "Considerations in Gravel Pack Design," *JPT* (Feb. 1974) 205-12; *Trans.*, AIME, **257.**
10. Holman, G.B.: "Evaluation of Control Techniques for Unconsolidated Silty Sands," *JPT* (Sept. 1976) 979-84.
11. Monroe, S.A. and Penberthy, W.L. Jr.: "Gravel Packing High Volume Water Supply Wells," *JPT* (Dec. 1980) 2097-2102.
12. McLeod, H.O. Jr. and Crawford, H.R.: "Gravel Packing for High-Rate Completions," paper SPE 11008 presented at the 1982 SPE Annual Technical Conference and Exhibition, New Orleans, Sept. 26-29.
13. Penberthy, W.L. Jr. and Cope, B.J.: "Design and Productivity of Gravel-Packed Completions," *JPT* (Oct. 1980) 1679-86.
14. Bell, W.T.: "Perforating Techniques for Maximizing Well Productivity," paper SPE 10033 presented at the 1982 SPE Intl. Petroleum Exhibition and Technical Symposium, Beijing, March 18-26.
15. Harris, M.H.: "The Effect of Perforating on Well Productivity," *JPT* (April 1966) 518-28; *Trans.*, AIME, **237.**
16. McLeod, H.O. Jr.: "The Effect of Perforating Conditions on Well Performance," *JPT* (Jan. 1983) 31-39.
17. Crouch, E.C. and Pack, K.J.: "Systems Analysis Use for the Design and Evaluation of High-Rate Gas Wells," paper SPE 9424 presented at the 1980 SPE Annual Technical Conference and Exhibition, Dallas, Sept. 21-24.
18. Mach, J., Proano, E.A., and Brown, K.E.: "Application of Production Systems Analysis to Determine Completion Sensitivity on Gas Well Production," paper 81-Pet-13 presented at the 1981 ASME Energy Sources Technical Conference, Houston, Jan. 18-21.
19. Cooke, C.E. Jr.: "Conductivity of Fracture Proppants in Multiple Layers," *JPT* (Sept. 1973) 1101-07; *Trans.*, AIME, **255.**
20. Katz, D.L. *et al.*: *Handbook of Natural Gas Engineering*, McGraw-Hill Book Co. Inc., New York City (1959) 47-50.
21. Muecke, T.W.: "Formation Fines and Factors Controlling Their Movement in Porous Media," *JPT* (Feb. 1979) 144-50.
22. Gruesbeck, C. and Collins, R.E.: "Entrainment and Deposition of Fine Particles in Porous Media," *SPEJ* (Dec. 1982) 847-56.
23. Gabriel, G.A. and Inamdar, G.R.: "Experimental Investigation of Fines Migration in Porous Media," paper SPE 12168 presented at the 1983 SPE Annual Technical Conference and Exhibition, San Francisco, Oct. 5-8.

Chapter 4
Gravel-Pack Design

Introduction

It is important to distinguish between the production of load-bearing solids and formation fines. Formation fines also are associated with a formation but may not be part of its mechanical structure. Fines are very small particles of loose solid materials present in the pore spaces of all sandstone reservoirs. They were incorporated into the formation over geologic time. Indications are that formation fines are always produced as a result of fluid flow (see **Fig. 4.1**), which is beneficial. If fines are able to move, and are carried through the gravel pack by the produced fluids, they will not concentrate at pore restrictions and cause plugging or reduce the pack permeability. The extent to which fines move through reservoir rock is not clear. In gravel packs, however, the fines are thought to originate at the interface between the gravel pack and the near-well formation, rather than at distant places in the reservoir. The higher flow velocities near the well probably contribute to increased fines mobility in this region.

The basic problem is controlling formation sand without excessively reducing well productivity. Effective control requires a good gravel-pack design and execution, including obtaining a representative sample of the formation sand; analyzing the formation grain-size distribution, selecting an optimum gravel size in relation to formation sand size to control formation sand movement, and using the optimum screen slot width to retain the gravel. Other items that affect the gravel pack are proper well preparation procedures, choice of an effective placement technique, and implementation of procedures that will not impair productivity. These items are discussed in Chaps. 6 through 8.

Formation Sampling and Sieving

Sampling. The first step in the gravel-pack design process is to obtain representative samples of the formation's sand and to design the gravel pack on the basis of the smallest representative sand stratum.[1] The grain-size distribution often varies through a particular sand body, and usually from one zone to another. To ensure representative measurements, therefore, a number of samples are usually needed.

Samples from closely spaced rubber-sleeve or conventional cores provide the best material on which to base the design.[2-6] Sidewall samples are the next-best source. Bailed samples, or samples obtained from separators, can also be used but may not be representative of formations because of size segregation of the sand particles. Information from offset wells should be used in the absence of other data. In any event, the sample sieved should be representative of the formation to be gravel packed.

Sieving. In gravel packing, formation sand is controlled by properly sized gravel. For correct gravel sizing, the producing formation's grain size must be determined accurately. The most widely used methods for assessing grain size are based on the standard screen scales, which grade the screen or sieve sizes in mesh numbers. In this technique, representative formation sand samples are extracted, dried, weighed, and passed through screens of various sizes.

The size and distribution of the formation sand, as determined from a sieved sample of the sand formation, are important to the optimal design of gravel packs. In practice, sieving is simple.[7] The disaggregated material to be sized is placed in a sieve stack and shaken until those particles smaller than the sieve openings fall through to the next-smallest sieve and so on. Separation into any number of size groups is possible if a progression of sieve sizes is used, from larger to smaller, downward from the top screen. Shaking should be done on an automatic machine for about 20 minutes (see **Fig. 4.2**). This length of time represents a compromise between obtaining complete grain-size separation and avoiding sediment-particle and screen erosion.

Sieving can be performed with dry, wet, or sonic techniques. In dry sieving, all moisture must first be removed. Wet sieving, often used on formation sands having a high clay content, involves flowing water over the screen in addition to shaking. This technique tends to separate the sand grains bound together by clay; the result is a representative sieve analysis. Regardless of the technique used, sieving must separate the formation sand into individual grains, and sufficient sieving time must be allowed for the sand grains to settle to the correct sieve size in relation to all particle sizes.

The theory of sieving is not quite as simple as the practice. Because most grains are not spherical, separation is not based solely on grain diameter. Grain shape also plays a part. A long, lathe-shaped grain theoretically could pass through a given sieve if its two smaller dimensions are less than those of the sieve opening. Despite this shortcoming, sieving is an established technique in the mechanical analysis of sediments.

Sieve analysis requires the use of a grade scale—an arbitrary, but systematic, division of sizes along a continuous scale into convenient class intervals. Several grade scales have been used. Some are arithmetic progressions; others are geometric. A geometric progression requires a fixed ratio between successive elements, such as the series 4, 2, 1, ½, ¼, ⅛, where each number is half the preceding one. This is the system followed in commonly used U.S. mesh series.

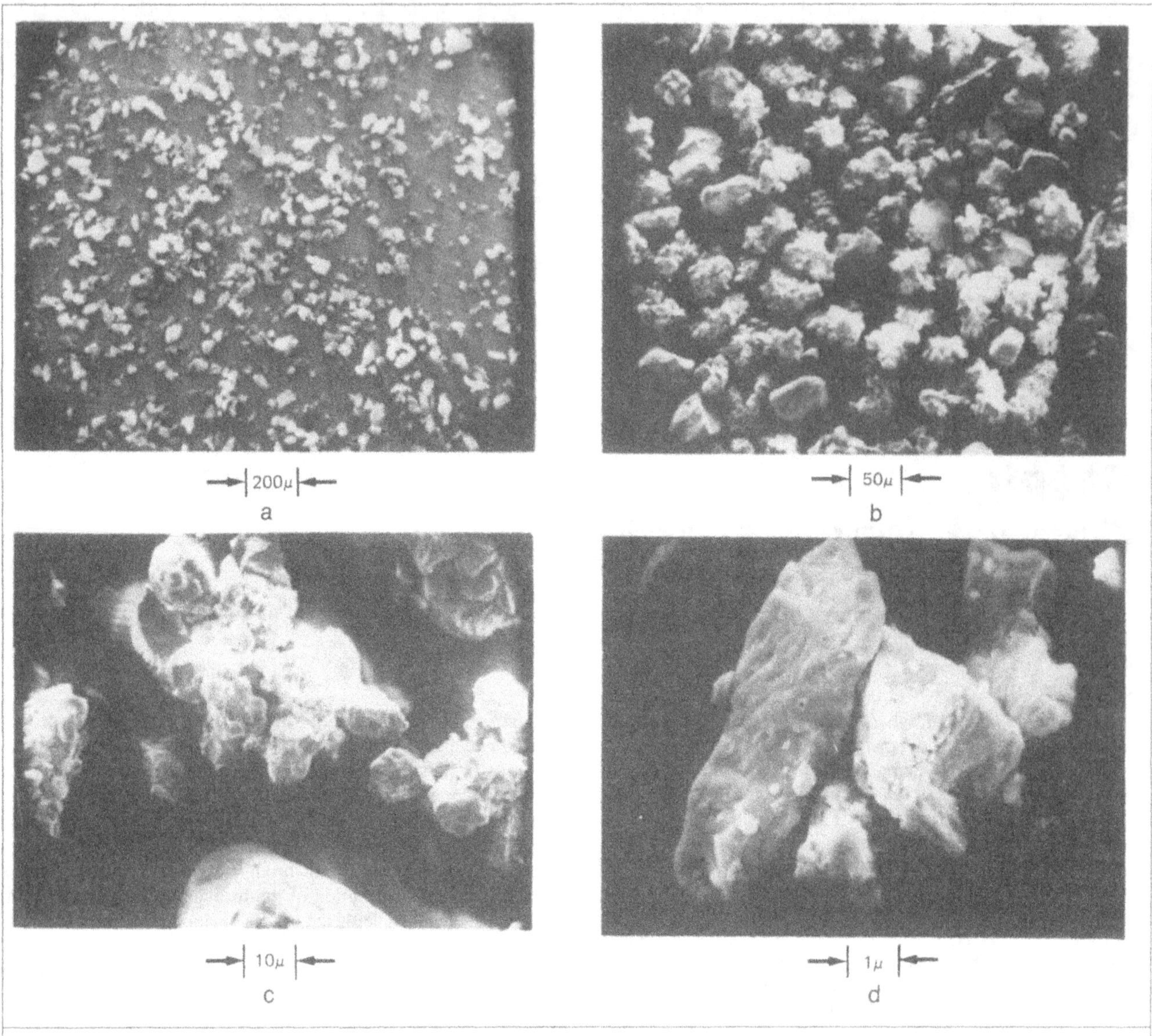

Fig. 4.1—Formation fines (< 400 mesh): (a) low magnification; (b) intermediate magnification; (c,d) high magnification.

The U.S. mesh series is based on a geometric progression of two. Hence, the area of the openings of each successively larger screen is increased by a factor of two and has twice the cross-sectional area (see **Fig. 4.3**). The diameter of grains retained on these openings thus increases geometrically by a factor of the square root of two, or 1.414. Closer grading can be achieved by including screens with openings with a progression of the fourth root of two, or 1.189. By themselves, these screens provide an offset scale that also increases by the square-root-of-two progression (see **Table 4.1**).

Table 4.2 compares sieve sizes, and **Table 4.3** lists the commonly available gravel sizes. The millimeter equivalents of these scales do not agree precisely with the Wentworth grade limits, but they are close enough to be within the U.S. Bureau of Standards limits of sieve tolerance.

The square-root-of-two progression yields four U.S. mesh sieves per Wentworth grade. **Fig. 4.4** shows the relationship among the Wentworth particle classification, the actual particle sizes, and the U.S. mesh series for a range that should cover most reservoir sediments. Other methods for classifying particle sizes are included for comparison.

After shaking, the material retained on each sieve and the particles that fell through the finest sieve into the pan are weighed and recorded. A cumulative weight-percent distribution as a function of grain size is plotted for use in designing the gravel pack. **Fig. 4.5** illustrates typical sieve analyses for several different formation sands.

The tabular classification of data does not lend itself to visual comparison of grain-size distributions. Comparisons are easier when a graph of grain size (horizontal scale) is used. The use of standard statistical devices has been modified because of the peculiarities of the graphic presentation of sieve data, which are based on weight instead of number of particles and a reversed logarithmic scale of diameter. **Figs. 4.6 through 4.8** are graphs of sample sieve analyses for various size distributions.

Three fundamental measures of the frequency diagrams are easily obtained from sieve data. The measures of central tendencies are the averages, the median, and the mean. The measure of variation in the distribution, the standard deviation, is generally called sorting. Skewness is a measure of asymmetry in the distribution.

Comparison and interpretation of the cumulative size/frequency distribution are fairly simple. If a curve maintains its form, a translation along the abscissa reflects a change in grain size (Fig. 4.6). A change in slope or steepness in the cumulative curve represents a change in the sorting of grain sizes (Fig. 4.7). Skewed distributions are asymmetrical, as shown in Fig. 4.8.

Development of Design Criteria

After representative samples of the formation sand are collected and sieved, the weights of the sieved formation sand recovered from each sieve are recorded and a sieve-analysis curve is constructed (Fig. 4.5). The slope of this sieve-analysis curve expresses the sand's uniformity. Curve A in Fig. 4.5, which is almost vertical, represents a highly uniform sand. But Curve D, which has a lower slope, suggests a nonuniform sand because of the wide distribution in particle

Fig. 4.2—Screening device through which sand is passed for measurement.

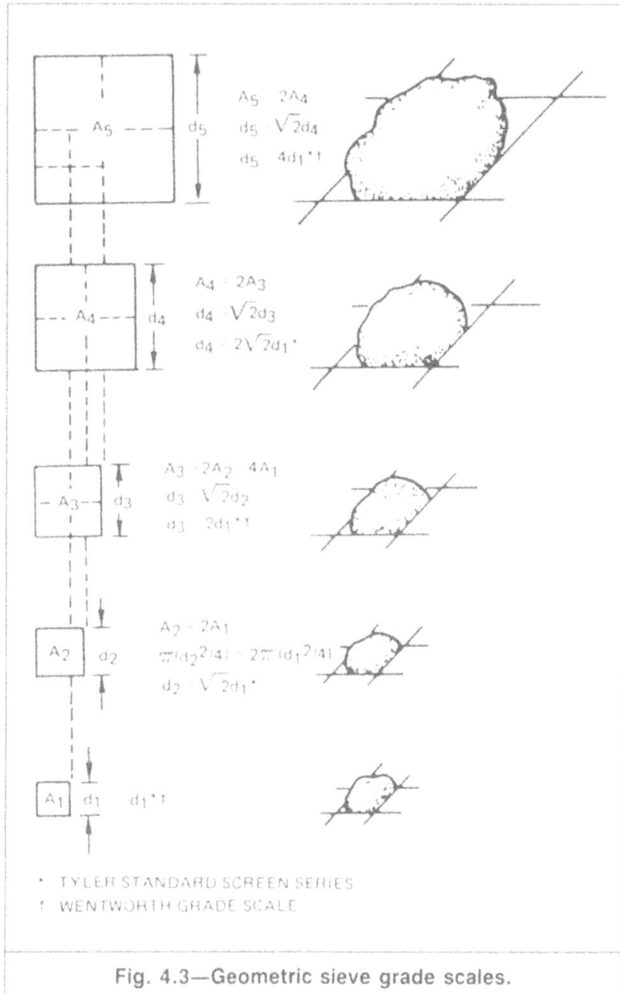

Fig. 4.3—Geometric sieve grade scales.

sizes compared with Curve A. A specific value of uniformity is useful in providing an approximate description of the distribution curve.

The convention has been to describe the distribution curve by the grain size at two specific percentile points, 40 and 90. The dimension of the grain size at the 40th percentile divided by that at the 90th percentile is called the uniformity coefficient, C_μ.[10]

$$C_\mu = d_{40}/d_{90}.$$

The uniformity coefficiency is a common basis for comparing sands. The utility of the uniformity coefficient in terms of arriving at a gravel-pack design is debatable; however, some design methods use it as a basis for selecting the design point. Except for its use to differentiate between sand samples, it may be academic. Although this relationship has become the accepted method for expressing uniformity, it is somewhat confusing: highly uniform sands have low uniformity coefficients and nonuniform sands have high coefficients. As a general guideline, for $C_\mu < 3$, the sand is considered uniform; for $3 < C_\mu < 5$, the sand is considered nonuniform; and for $C_\mu < 5$, the sand is highly nonuniform.

Design Point. A gravel pack usually should be designed to restrain the load-bearing grains of the formation sand from passing into the wellbore. These grains will bridge the formation sand, controlling the entry of formation sand into the well while permitting the passage of fluids. Normally, a specific grain size is selected as a basis for the gravel-pack design. This size is used as the design point for sizing gravel, because gravel size is a multiple of the formation sand size at the design point. Experience and available data show that the choice of design points has changed with time. Early packs were designed by using the 10th percentile point as the basis.[11-13] Many engineers now use a higher design point, usually the 50th percentile.[14,15] Some designs specify other design points; others

TABLE 4.1—STANDARD SCREEN SIEVES

Tyler Mesh or Tyler Number		
Progression on $(2)^{1/2}$	Progression $(2)^{1/4}$	Opening Diameter (in.)
3½	3½	0.2210
4	4	0.1850
	5	0.1560
6	6	0.1310
	7	0.1100
8	8	0.0930
	9	0.0780
10	10	0.0650
	12	0.0550
14	14	0.0460
	16	0.0390
20	20	0.0328
	24	0.0276
28	28	0.0232
	32	0.0195
35	35	0.0164
	42	0.0138
48	48	0.0116
	60	0.0097
65	65	0.0082
	80	0.0069
100	100	0.0058
	115	0.0049
150	150	0.0041
	170	0.0035
200	200	0.0029
	250	0.0024
270	270	0.0021
	325	0.0017
400	400	0.0015

TABLE 4.2—STANDARD SIEVE SIZES

U.S. Mesh		Sieve Opening	
U.S. Series	Tyler Series	in.	mm
2½		0.315	8.00
	2½	0.312	7.925
3		0.265	6.73
	3	0.263	6.68
3½		0.223	5.66
	3½	0.221	5.613
4		0.187	4.76
	4	0.185	4.699
5		0.157	4.00
	5	0.156	3.962
6		0.132	3.36
	6	0.131	3.327
7		0.111	2.83
	7	0.110	2.794
8		0.0937	2.38
	8	0.093	2.362
10		0.0787	2.00
	9	0.078	1.981
12		0.0661	1.68
	10	0.065	1.651
14		0.0555	1.41
	12	0.055	1.397
16		0.0469	1.19
	14	0.046	1.168
18		0.0394	1.00
	16	0.0390	0.991
20		0.0331	0.84
	20	0.0328	0.833
25		0.0280	0.71
	24	0.0276	0.701
30	28	0.0232	0.589
35		0.0197	0.50
	32	0.0195	0.495
40		0.0165	0.42
	35	0.0164	0.417
45	42	0.0138	0.351
50		0.0117	0.297
	48	0.0116	0.295
60		0.0098	0.250
	60	0.0097	0.246
70		0.0083	0.210
	65	0.0082	0.208
80		0.0070	0.177
	80	0.0069	0.175
100		0.0059	0.149
	100	0.0058	0.147
120	115	0.0049	0.124
140	150	0.0041	0.104
170	170	0.0035	0.088
200	200	0.0029	0.074
230	250	0.0024	0.062
270	270	0.0021	0.053
325	325	0.0017	0.044
400	400	0.0015	0.037

				Permeability and Porosity of Graded Sands[9]		
U.S. Mesh	8/12 Angular	10/20 Angular	10/20 Round	10/30 Round	20/40 Round	40/60 Round
Permeability, darcies (approximately)	1,745	881	325	191	121	45
Porosity, % (approximately)	36	36	32	33	35	32

TABLE 4.3—COMMONLY AVAILABLE GRAVEL SIZES

Gravel Size (in.)	U.S. Mesh Size	Approximate Median Diameter in.	Approximate Median Diameter μm
0.006 x 0.017	40/100	0.012	300
0.008 x 0.017	40/70	0.013	330
0.010 x 0.017	40/60	0.014	350
0.017 x 0.033	20/40	0.025	630
0.023 x 0.047	16/30	0.035	880
0.033 x 0.066	12/20	0.050	1260
0.039 x 0.066	12/18	0.053	1340
0.033 x 0.079	10/20	0.056	1410
0.047 x 0.079	10/16	0.063	1590
0.066 x 0.094	8/12	0.080	2020
0.079 x 0.132	6/10	0.106	2670

specify a design point that is dependent on the uniformity coefficient.[16,17] The currently used design points are listed below and are expressed as a multiple of the formation and gravel sizes. That is, D50G is the diameter at the 50th percentile point of the gravel, while D50F is the diameter of the 50th percentile point for the formation.

D50G = 6(D50F) (Saucier[15]).

D10G = 6(D10F) for $C_\mu < 5$,

D40G = 6(D40F) for $C_\mu > 5$, and

D70G = 6(D70F) for $C_\mu > 10$ (Schwartz[17]).

D85G ≤ 4(D15F) (Stein[16]).

D10G ≤ 10(D10F) (Coberly and Wagner[12]).

Of these choices, the 50th percentile point currently seems to be common in the U.S. gulf coast, but geographical and other factors may indicate another design point. If the sands are uniform, all gravel-pack designs tend to be the same, regardless of the sizing method used, because there is little difference in grain sizes between the 10th, 50th, and 70th percentile points. For nonuniform sands, however, the designs will yield different gravel sizes, and a considerable difference in gravel sizes will exist. Hence, the selection of the appropriate design point becomes more critical for nonuniform sands. Designing a gravel pack on the 50th percentile point represents a compromise for most conditions. It will yield about the same design for a uniform sand that the 10th percentile would and will provide the optimum design for nonuniform sands. In some gravel packs in certain locations, other design points yielded optimum results. Consequently, the selection of an appropriate design point should be based on the design that yields the best field results.

Data indicate that a gravel pack should not be designed to block all formation particles, which occurs when high design points are

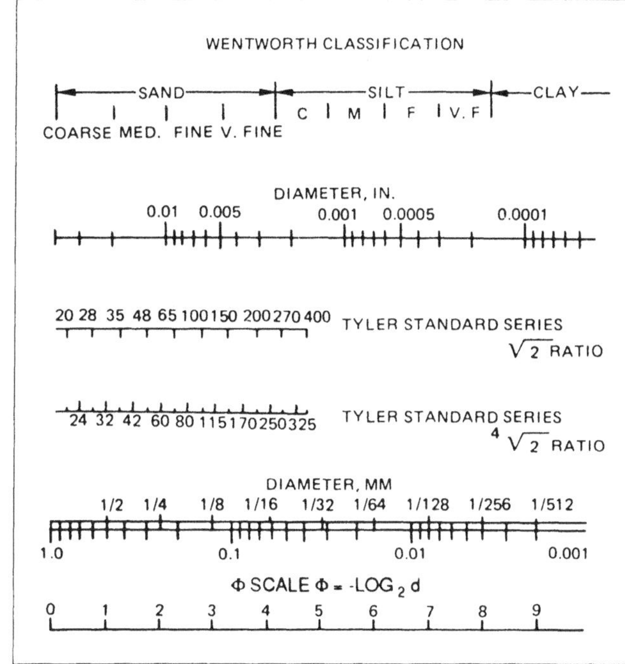

Fig. 4.4—Relationship between particle size and standard screens.

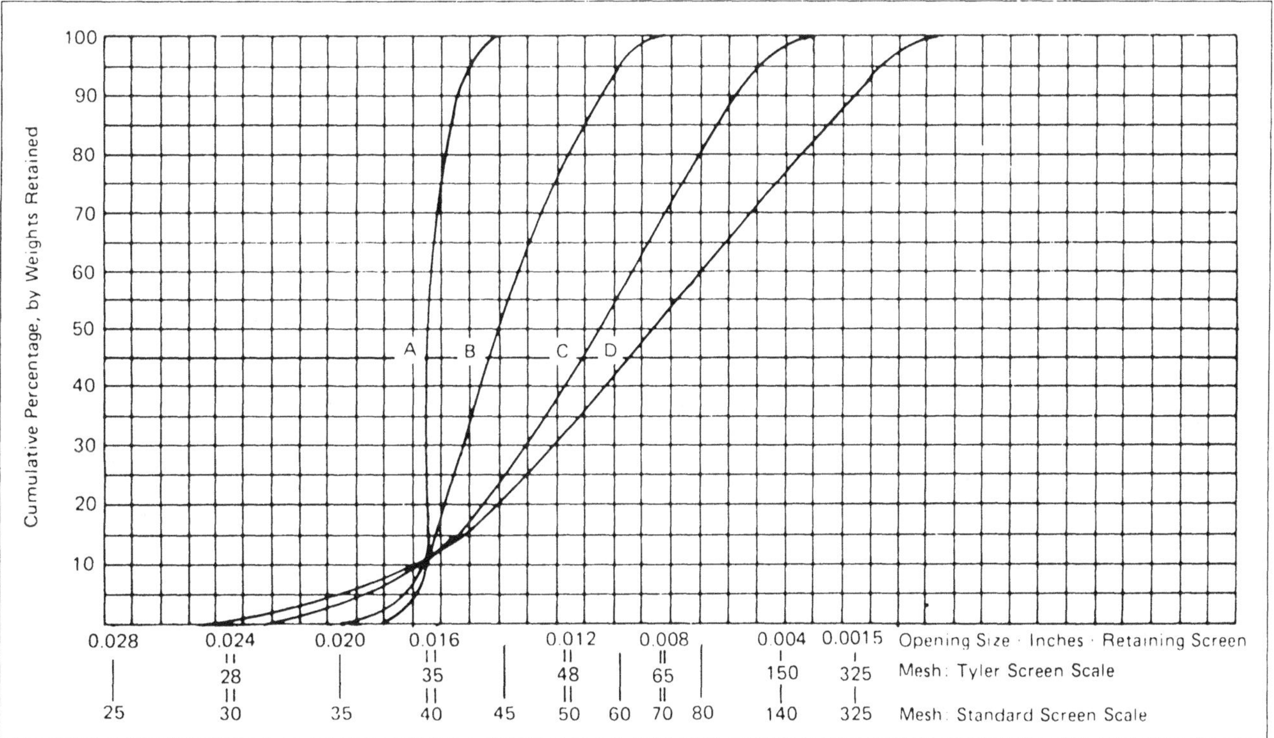

Fig. 4.5—Sieve-analysis curves constructed from the cumulative percentage of sand retained after processing through the device shown in Fig. 4.2 vs. grain size.

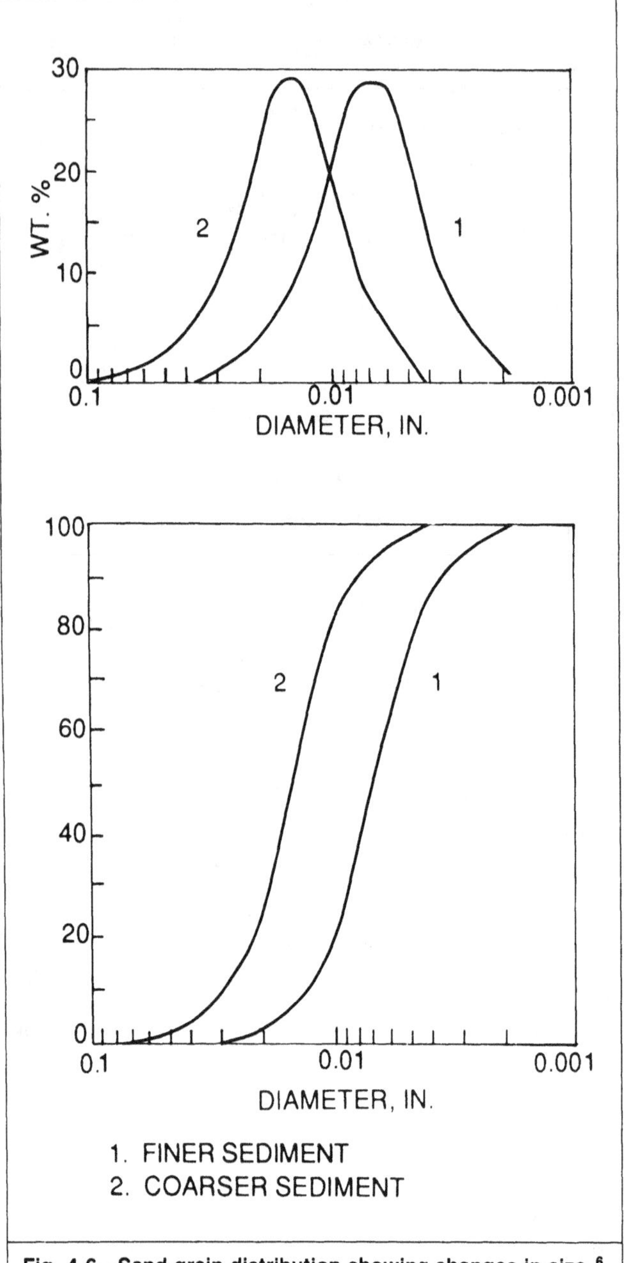

Fig. 4.6—Sand-grain distribution showing changes in size.[6]

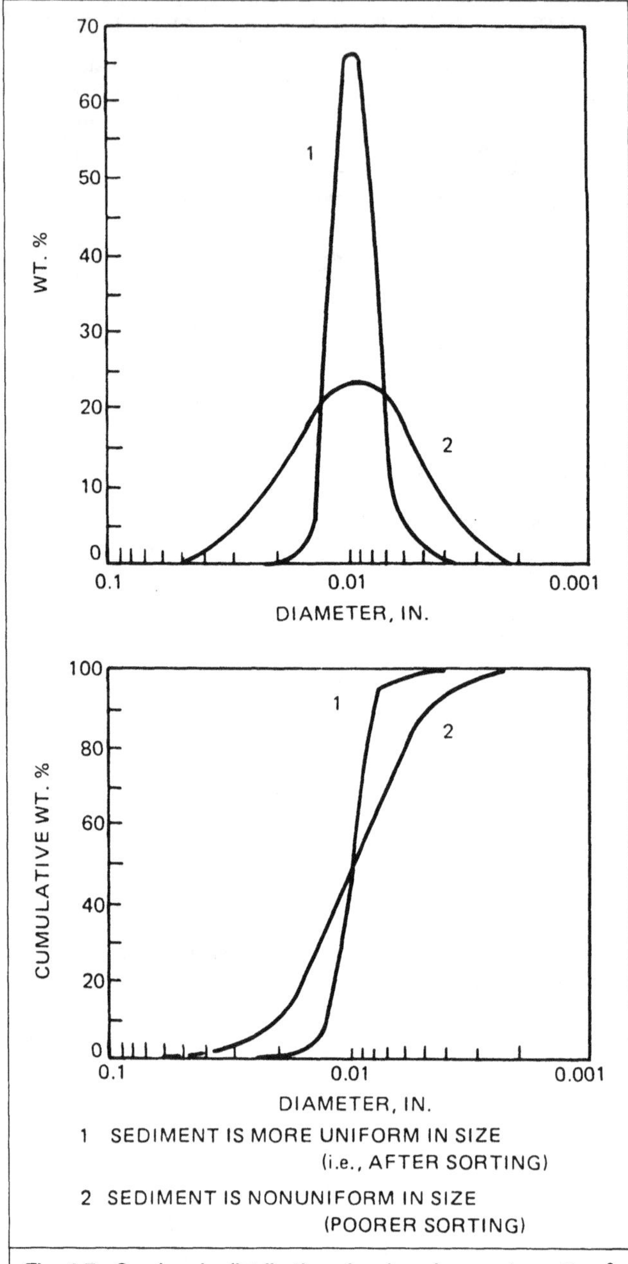

Fig. 4.7—Sand-grain distribution showing changes in sorting.[6]

used.[14] If high design points are chosen, well productivity can be severely limited by total plugging of the gravel pack by fine particles. Again, this material usually is produced through the gravel pack if it is designed to retain only the load-bearing particles.

Problems arise when nonuniform sands are to be gravel packed because the range of particle sizes that must bridge is large. Under these conditions, no single design point accurately describes the entire formation and represents a uniform sand. The result can be a gravel pack with good sand control, but at the expense of productivity.

A gravel-pack design based on the uniformity coefficient criterion can be greatly influenced by the way that the formation sand sample is collected. If most of the finer-grained material is lost, the sample's uniformity coefficient may register lower than it actually is, resulting in the selection of oversized gravel. Regardless of the design point used, this example demonstrates the need for a representative sample and proper analysis.

Gravel/Sand Ratio. The gravel/sand ratio—the ratio of the gravel grain size to the formation sand grain size at equal percentile points—is one of the most important parameters in a gravel-pack design. When the gravel/sand ratio is high, the oversized gravel is invaded by formation sand, which reduces the overall permeability of the packed zone (often to less than the native reservoir permeability) and impairs productivity. On the other hand, using undersized gravel results in excellent sand control but, in certain situations, may jeopardize productivity.

The theoretically optimum range[14,15,17] for the gravel/sand ratio is about 5 to 6, as illustrated in **Fig. 4.9.** Note that permeability is maximum when the gravel/sand ratios are less than or greater than 10. At a ratio of about 6, the gravel-pack permeability is at a maximum and sand is also being controlled. By contrast, at a ratio of 15, pack permeability is good but sand control is poor because the formation sand can move easily through the pack. At a ratio of 10, formation sand can move into the gravel pack but has difficulty moving through it. The result is a severe loss in gravel-pack productivity.

To compensate for possible optimism and to adjust for an erroneous choice of design point (caused, perhaps, by improper sampling), a gravel/sand ratio of 5 to 6 is commonly used. That is, the diameter of the gravel to be used for packing is five to six times the diameter of the formation sand at the 50th percentile design point.

Given the sieve analysis in **Fig. 4.10,** the procedure for designing a gravel pack based on the 50th percentile design point and a gravel/sand size ratio of 6 would be as follows.

1. Find the 50th percentile point.
2. Multiply this value by six.

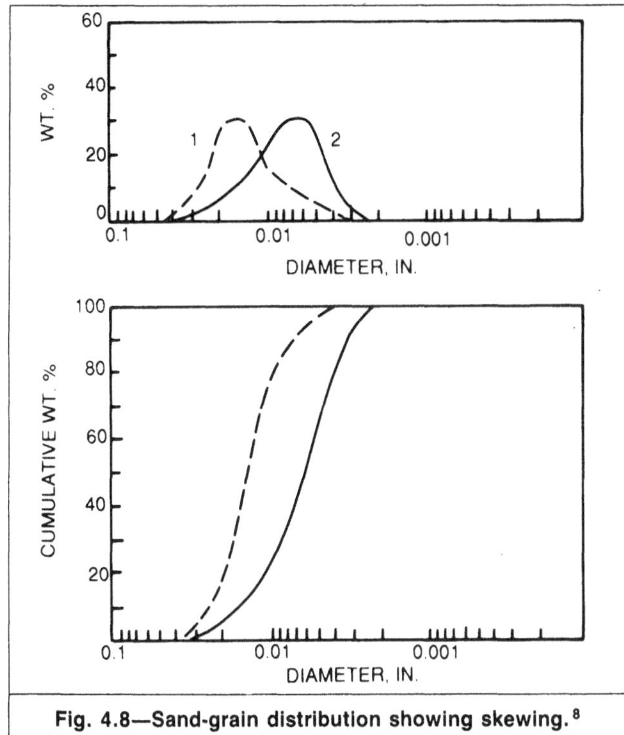

Fig. 4.8—Sand-grain distribution showing skewing.[8]

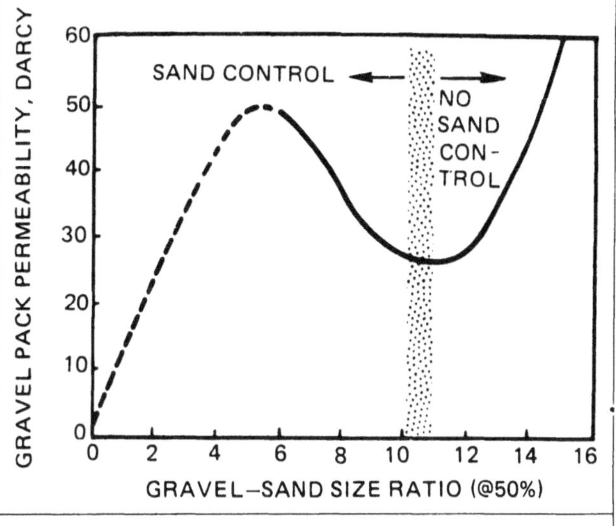

Fig. 4.9—Effect of gravel/sand size ratio on sand control and productivity.

3. Construct a gravel curve through this point—i.e., $C_\mu = 1.5$, $d_{40}/d_{90} = 1.5$.

4. Select the gravel size from the 0th and 100th percentile points (choose the closest standard gravel size at each point).

5. Select a screen slot width one-half the smallest gravel size.

Fig. 4.10 shows that when the 50th percentile point is multiplied by six, that point on the gravel curve is 0.024 in. The gravel curve is then constructed so that its uniformity coefficient is 1.5. The gravel size is taken from the 0th and 100th percentile points (to the closest standard size for each). The screen-slot size is one-half the smallest gravel size. For this example, a 20/40-U.S.-mesh gravel was selected. Because 40-U.S.-mesh gravel has a diameter of 0.0165 in., the slot size should be 0.008 to 0.010 in.

Other gravel-pack designs are possible by selection of different design points and gravel/sand ratios.

Screen-Slot Width. Ideally, slots should be as wide as possible while retaining sand grains without restricting flow of fluids and interstitial fines. Because all the gravel must be tightly packed and retained, the screen-slot width for a gravel pack should be about one-half the smallest gravel diameter. The slot should not be wider than 70% of the smallest gravel-size diameter to avoid the production of gravel. Theoretically, a slot width greater than 70% of the smallest gravel size will retain it. However, screen manufacturing tolerances may vary, gravel sizes may not be exact, and the screen could be damaged during the completion or eroded during production. Because the flow capacities of screens and slotted liners are high, sizing the slot width on the low side will not restrict productivity and will tend to ensure that the gravel remains outside the screen.

Gravel-Pack Thickness. Many opinions have been expressed regarding the optimum thickness of gravel packs. Carefully controlled laboratory experiments have shown that a thickness of only 3 to 5 grain diameters is required to create a stable bridge.[11] This conclusion, however, is based on the assumption that the bridge will

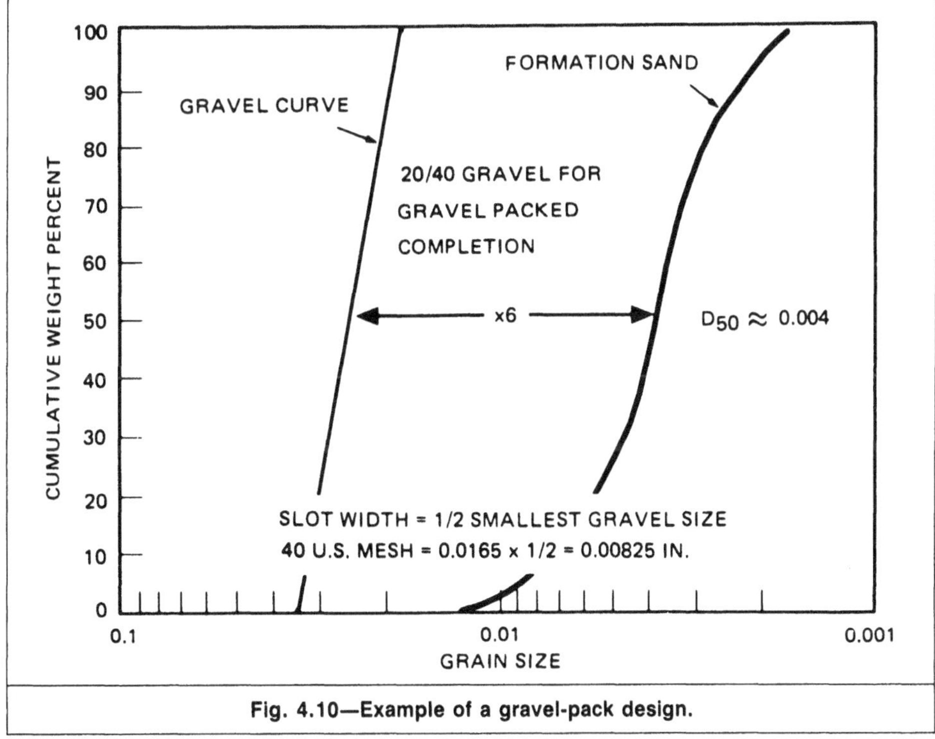

Fig. 4.10—Example of a gravel-pack design.

be permanent and completely effective. Occasional adjustments of sand particles with changes in differential pressures and flow velocities will no doubt occur. With each failure of a sand bridge, individual grains—particularly the finer grains—may encroach further into the gravel pack. It follows that the thicker the gravel bed, the greater the assurance that the pore spaces between gravel particles will remain unclogged with sand. Gravel-pack design should therefore place more emphasis on pack thickness than on screen diameter. However, attention must be given to well operability. Tools will have to enter the gravel-packed interval and screen removal may be necessary on occasion. Both of these requirements affect the gravel-pack thickness.

Operators should be able to run the screen in the well without problems, remove it easily, and wash over it. The screen diameter should also be large enough that wireline tools, production logs, or pumps can be run inside it if needed. When all variables are considered, the following guidelines are offered for screen diameters inside various casing sizes. Hence, the clearance sets the thickness of the gravel pack.

Casing Size (in.)	Screen Size (in.)
9⅝	5
7⅝	3½
7	2⅞ to 3½
5½	2⅜
4	1¼ to 1½

Gravel Specifications. Gravel that meets critical specifications can also affect completion success. Regardless of the gravel size required for a particular completion, recommendations are that 96 wt% of it be within the specified range. No more than 2% should be finer than that specified.

In addition to the size specification, the gravel should be rounded at its edges and spherical to minimize fines generation during handling and to provide low pack porosities. A roundness and sphericity of >0.6 are suggested. The crush resistance of the gravel should be such that no more than 5 wt% fines is generated at a stress of 2,000 psi. The solubility in 12/3% (hydrochloric and hydrofluoric acid) should not exceed 1% when exposed for 1 hour. The low-fines-content requirement of the gravel can be met by use of triple-sieving techniques offered by suppliers to meet these gravel requirements. Recommended practices for testing gravel used in gravel-packing operations are discussed in detail in API *RP 58*.

While most wells are gravel-packed with silica gravel particles, other materials can be used if they meet the above specifications. Sintered bauxite and similar intermediate-strength materials have been used for gravel packing, but not as frequently as sieved gravel. They differ from ordinary gravel in that they have much higher crush resistances and higher densities. Another material available is glass beads; their use should be limited, however, to situations where no hydrofluoric acidizing is anticipated because the beads are highly acid-soluble.

Summary

To achieve optimum results from gravel packing, the operator should do the following.

1. Obtain a representative sample of formation sand. Rubber-sleeve and conventional cores are best; sidewall cores are acceptable; bailed or produced sand should be used only as a last resort. Use offset-well data in the absence of specific well information.
2. Sieve the formation sand and plot a cumulative weight distribution. Calculating the coefficient of uniformity may be useful—i.e., $C_\mu = D_{40}/D_{90}$.
3. Select an appropriate design point on the sieve-analysis curve.
4. Select the gravel diameter by multiplying the design point by the gravel/sand size ratio (in the range of 4 to 8 times the design point size).
5. Specify the narrowest range of sieve sizes that would successfully contain the selected gravel diameter. The gravel curve should be constructed so that its uniformity coefficient is 1.5.
6. Use a screen-slot size that is one-half the smallest gravel size. In any case, slot size should not exceed 70% of the smallest gravel diameter.
7. Use recommended screen diameters for various casing sizes. Because screen flow capacities are extremely high, the screen-diameter recommendation is based on a diameter that allows the entry of wireline tools and other production equipment and allows it to be washed over.

References

1. Maly, G.P. and Krueger, R.F.: "Improper Formation Sampling Leads to Improper Selection of Gravel Size," *JPT* (Dec. 1971) 1403–08.
2. Jennings, H.Y.: "How To Handle and Process Soft and Unconsolidated Cores," *World Oil* (June 1965) 116–18.
3. "New Sleeved Core Barrel Recovers Soft Formation," *World Oil* (Feb. 1, 1971) 40.
4. Hildebrandt, A.B., Birdwell, H.C., and Kellner, J.M.: "Development and Field Testing of a Core Barrel for Recovering Unconsolidated Oil Sands," *Trans.*, AIME (1958) **213,** 347–49.
5. Austin, B.L.: "Rubber Sleeve Core Barrel Increases Recovery," *World Oil* (May 1959) 159–61.
6. Rogers, E.B. Jr.: "Sand Control in Oil and Gas Wells," *Oil & Gas J.* (Nov. 1, 8, 15, and 22, 1971) 54–68.
7. Knutson, C.F., Sutton, E.W., and Cavanaugh, R.J.: "New Technique Increases Core Recovery," *World Oil* (Feb. 1, 1961) 37–40.
8. Perry, R.H.: *Chemical Engineers' Handbook*, sixth edition, McGraw-Hill Book Co. Inc., New York City (1984).
9. van Poollen H.J., Tinsley, J.M., Saunders, C.D.: "Hydraulic Fracturing—Fracture Flow Capacity vs Well Productivity," *Trans.*, AIME (1958) **213,** 91–95.
10. Krumbein, W.C. and Pettijohn, F.J.: *Manual of Sedimentary Petrography*, Appleton-Century Co., New York City (1938).
11. Coberly, C.J.: "Selection of Screen Openings for Unconsolidated Sands," *Drill. & Prod. Prac.*, API, Dallas (1937) 189–201.
12. Coberly, C.J. and Wagner, E.M.: "Some Considerations in the Selection and Installation of Gravel Pack for Oil Wells," *Pet. Tech.* (Aug. 1938) AIME Tech. Publ. No. 960, 1–20.
13. Hill, K.E.: "Factors Affecting the Use of Gravel in Oil Wells," *Drill. & Prod. Prac.*, API, Dallas (1941) 134–43.
14. Penberthy, W.L. Jr. and Cope, B.J.: "Design and Productivity of Gravel-Packed Completions," *JPT* (Oct. 1980) 1679–86.
15. Saucier, R.J.: "Considerations in Gravel Pack Design," *JPT* (Feb. 1974) 205–12; *Trans.*, AIME, **257.**
16. Stein, N.: "Designing Gravel Packs for Changing Well Conditions," *World Oil* (Feb. 1, 1983) 41–47.
17. Schwartz, D.H.: "Successful Sand Control Design for High Rate Oil and Water Wells," *JPT* (Sept. 1969) 1193–98.

SI Metric Conversion Factor

in. × 2.54* E+00 = cm

*Conversion factor is exact.

Chapter 5
Screens and Slotted Liners

Introduction

In gravel packing, accurately sized gravel is used as a filtering medium. The gravel is retained in an annular region within the wellbore by a mechanical device (see **Fig. 5.1**), such as torch-cut slots, perforated casing, louver-type screen, machine-slotted pipe, and several variations of wire-wrapped screen.[1] In most current gas- and oil-well completions requiring gravel packing, however, either slotted liners or wire-wrapped screens are preferred.

The slotted liners usually have vertical slots spaced uniformly about the casing. The width of the slots can vary from 0.020 in. to as large as desired, depending on the gravel size and the particular situation. Manufacturing constraints usually prevent slot sizes smaller than 0.020 in. **Fig. 5.2** illustrates various slot combinations for slotted liners. While slotted liners are much less expensive than wire-wrapped screens, they do not have high-inlet-flow areas and thus are not as popular for high-rate wells. Their use is usually confined to situations where wire-wrapped screens cannot be used economically—typically long completion intervals or low-productivity wells.

Description of Equipment

Wire screens are available in three basic constructions.

Welded 1 or 2. In Welded 1, the wire is resistance welded to longitudinal rods on a lathe-type machine. The screen (jacket) is manufactured separately, then placed over a drilled pipe base and welded into position (**Fig. 5.3**). Welded 2 is a variation of Welded 1 fabrication. In this case, the screen is wrapped and resistance welded directly to the pipe base, with the longitudinal rods as the only separation. The resistance weld here is still at the points of contact between the wire and the longitudinal rods. This manufacturing technique eliminates excess clearance between the screen and pipe base.

Wire wrapped. In these screens, the wire is wrapped directly on the pipe base, which may be drilled, slotted, or slotted and grooved. Longitudinal rods are also used between the screen and the pipe in some cases but are not welded to the wire screen at the contact points. The screen-slot openings are controlled by lugs, which are crimped into the wire during screen fabrication. After the wire is wrapped on the pipe base, it is welded at each end of the screen joint and is held in place by two longitudinal welds or solder joints that extend the length of the screen.

These longitudinal welds or solder joints keep the screen intact because it is held on the pipe base only by the force of winding it and by the welds at the ends. The longitudinal welds also keep the screen from unwinding in the event of damage during the running or retrieving of tools. **Fig. 5.4** is a photograph of a wire-wrapped screen.

Rod base. These screens are similar to the jackets on welded construction screens but contain no pipe base. For these screens, the wire is resistance welded to longitudinal rods and connections are welded on the ends of each screen. **Fig. 5.5** is an example of a rod-base screen.

Welded 1 and 2 and wire-wrapped screens offer much more open area per linear foot than slotted liners do (see **Fig. 5.6**).* Consequently, they are less susceptible to plugging in spite of their small slot widths (0.003 to 0.004 in.). They are used in most completions instead of slotted liners if the additional cost can be justified. Other screen designs have been used to permit their installation in special applications. One of these is a screen design that permits the screen jacket to expand along the pipe base in response to changes in temperature that occur in thermal-recovery operations.

The wire on Welded 1 and 2, wire-wrapped, and rod-based screens is usually keystone in shape (Fig. 5.3)—i.e., it tapers toward the center. This design avoids the liability of plugging once particulates pass through the screen slots. Although this configuration is also available on slotted liners by undercutting the slots, it is uncommon. Unless otherwise specified, all wire is usually 304 stainless steel. The wire can be made of any specified material, but a stainless-type steel is most common.

This chapter relates the results of tests done on wire screens to determine their strength characteristics. The results of the study were compared with the results of the same tests performed on slotted pipe.

Screen/Slotted-Liner Tests

Wire screens and slotted liners have been tested for tensile strength, collapse resistance, and flow capacity.** Their flow capacities are discussed in Chap. 3. Screens from five U.S. manufacturers were tested.

Tensile-Strength Tests. All tensile-strength testing was performed on a standard testing machine routinely used for testing materials. This machine is designed so that a constant loading rate could be applied to the test sample until failure occurred. Sections of 2⅞- and 3½-in. Grade J-55 pipe-based screen or slotted pipe with standard 10-round nonupset male end connections were tested. One test

*Howard Smith Co. catalog (1982).

**Unpublished Exxon Production Research Co. report, Houston.

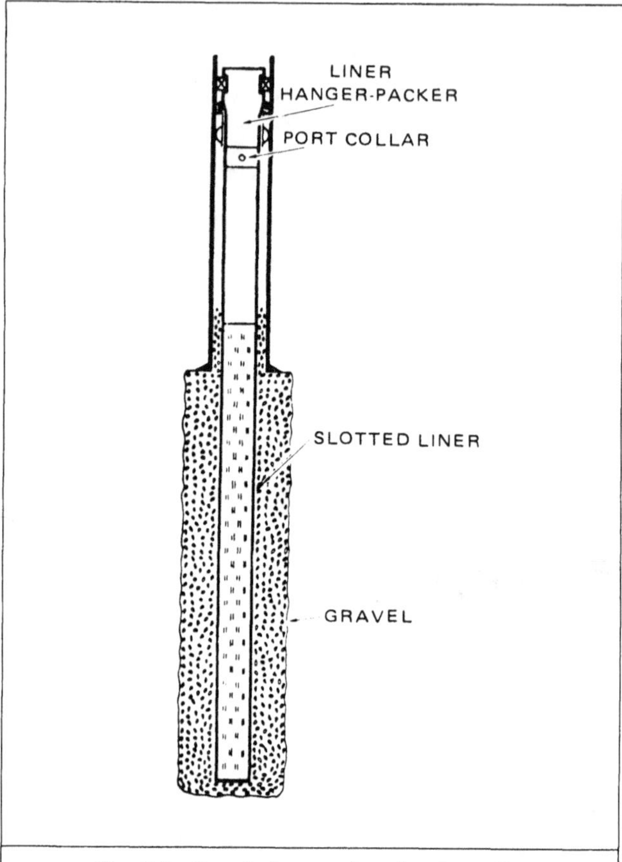

Fig. 5.1—Openhole gravel-pack schematic.

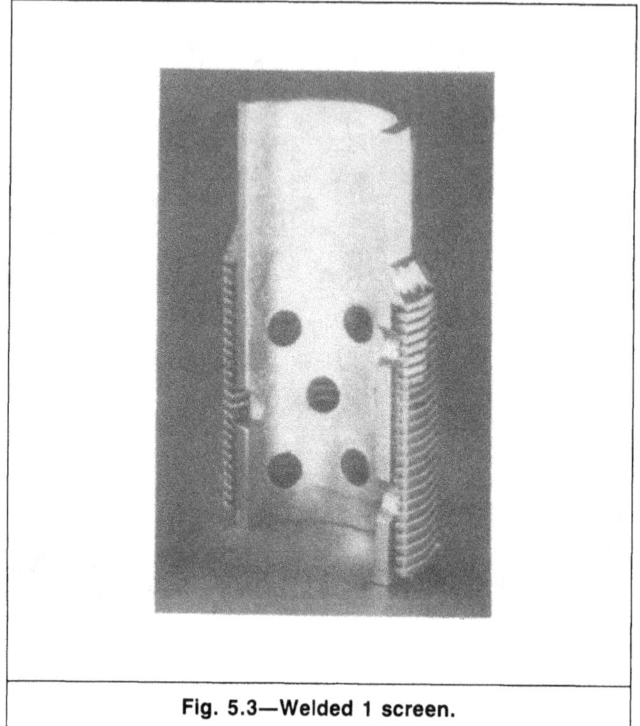

Fig. 5.3—Welded 1 screen.

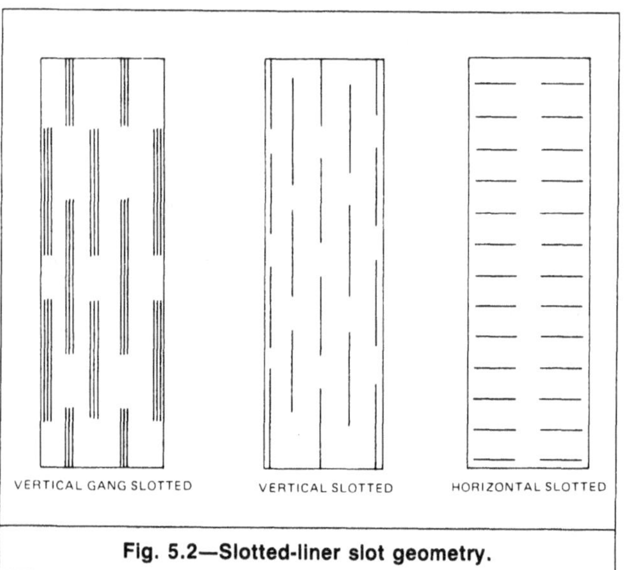

Fig. 5.2—Slotted-liner slot geometry.

Fig. 5.4—Wire-wrapped screen.

was made on a 1-in. screen. All the screens subjected to tensile-strength tests had 0.012-in. openings. The slotted pipe had 0.020-in. slots.

Collapse Tests. All collapse testing was conducted in the test cell shown in **Figs. 5.7 and 5.8.** It was designed so that the test sample could be inserted into the cell before hydrostatic pressure was applied to its circumference. Communication across the 0.008-in.-opening screen (0.020-in. slotted pipe) was prevented by wrapping the test sample with 1/16-in.-thick Buna N™ rubber sheeting and subsequently clamping the ends to the test sample.

Flow-Capacity Tests. Flow-capacity testing was conducted in the equipment illustrated in Fig. 3.5 to determine the pressure drop associated with increasing flow rate.

Test Results

Tensile-Strength Tests. Table 5.1 gives the results of the tensile-strength tests for the various screen constructions. The failures are recorded at the maximum load reached before test failure. On pipe-based screens, these loads were also converted to an equivalent stress based on the critical area of the sample, the cross-sectional area of the tubing under the last engaged thread.

Except for one test, coupling failure caused all pipe-based screen failures. The same failure sequence was observed in most cases: as load was applied to the test sample, the slot openings enlarged to the point that the longitudinal rods between screen and base separated from their welds at one of the sample ends. As additional load was applied, the test sample continued to elongate. Failure then occurred as a result of coupling (pin) pull-out caused by metal yield-

Fig. 5.5—Rod-based screen.

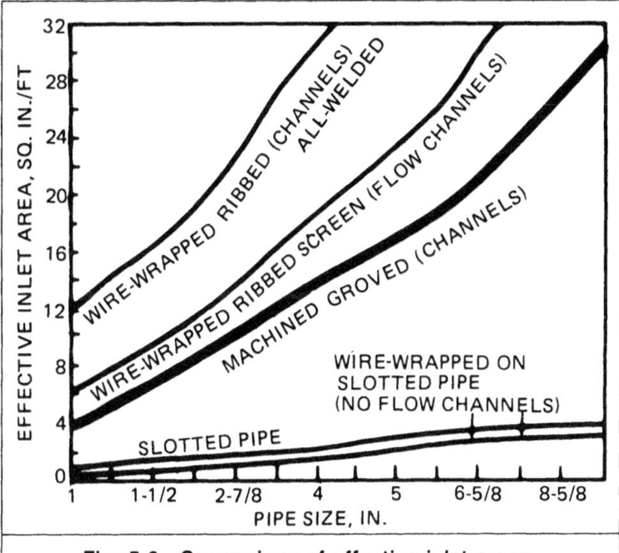

Fig. 5.6—Comparison of effective inlet areas.

ing. Screen rod separation usually occurred within 5,000 to 10,000 lbf of the coupling failure. **Fig. 5.9** is an example of a tensile failure.

Three observations can be made about screen yielding.

1. The ultimate failure stress always exceeded the published minimum yield strength.

2. The sample was elongated after the test, which verified that the yield strength had been exceeded.

3. Thread connections to the sample were tight before testing. After the test, however, the elongation was so extensive that the coupling could usually be unscrewed by hand from the end opposite the coupling failure. In the coupling failure in Fig. 5.9, the pin is elongated. Note also that the drilled pipe-base holes are elongated, further evidence of yielding.

The threads on the test samples were 10-round, nonupset API threads standard for welded screens used in the field. The minimum strength of this connection is equal to that of the pipe body (minimum yield of 55,000 psi). When yielding occurred, the threaded sections also yielded until the threads separated. That was the point of failure. The same results were noted on slotted pipe, where failure also involved separation at the coupling. Except for the 1-in. screen sample, no sample was pulled apart. This test sample was pulled apart under the last engaged thread, but the failure load was about five times the minimum joint yield strength.

These test results indicate that the drilled pipe-based construction of wire screens does not weaken the tensile strength of the screen

Fig. 5.7—Collapse test cell.

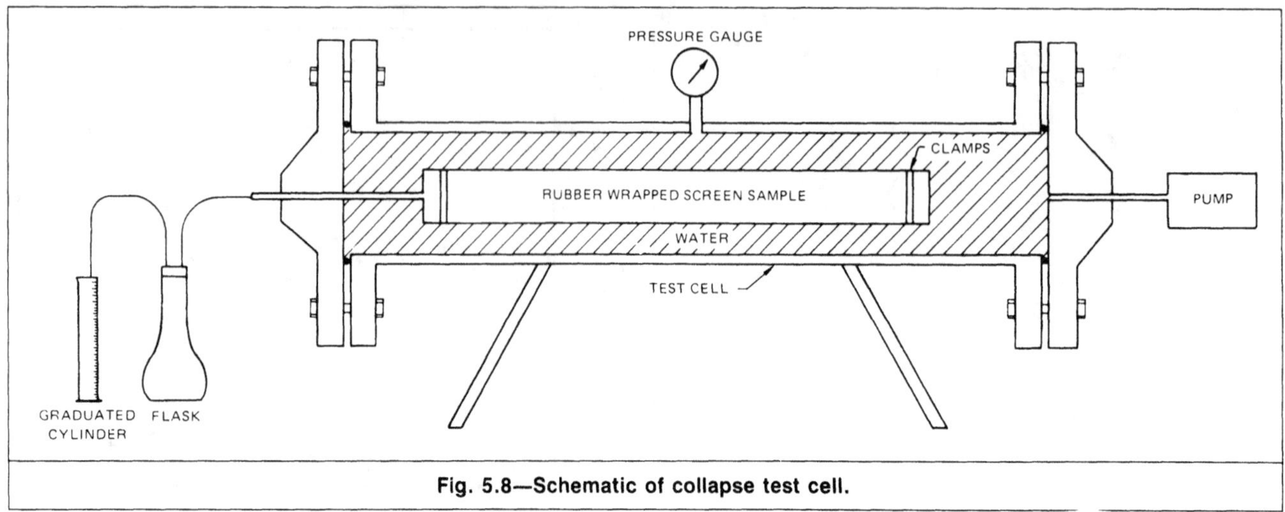

Fig. 5.8—Schematic of collapse test cell.

TABLE 5.1—SCREEN/SLOTTED-LINER TENSILE-STRENGTH TEST RESULTS

Screen/Slotted Liner Construction	Screens/Liners Tested	Screen/Liner Size (in.)	Average Failure Load (lbf)	Stress (psi)	Type Failure
Welded (1) screen	1	1	42,500	267,500	Threads failed
Welded (1) screen	5	2⅞	101,500	74,000	Coupling failed, screen elongated
Welded (1) screen	5	3½	147,000	77,000	Coupling failed, screen elongated
Wire-wrapped screen	2	2⅞	101,000	76,000	Coupling failed, screen elongated
Wire-wrapped screen	2	3½	149,000	75,000	Coupling failed, screen elongated
Rod-based screen	1	2⅞	48,000		Weld at coupling failed
	1	3½	70,000		Weld at coupling failed
Vertical slotted liner	1	2⅞	105,000	80,000	Coupling failed and elongation
Vertical gang slotted liner	1	2⅞	101,000	76,000	Coupling failed and elongation

Screen/Liner Size (in.)	ID (in.)
1	0.824
2⅞	2.441
3½	2.992

Fig. 5.9—Typical tensile failure of Welded 1 screen.

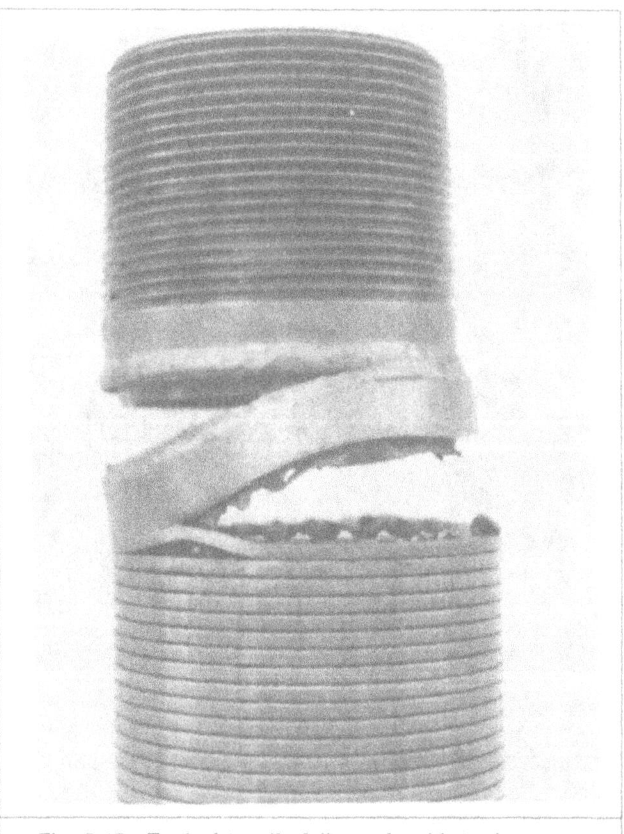

Fig. 5.10—Typical tensile failure of rod-based screen.

TABLE 5.2—SCREEN/SLOTTED-LINER COLLAPSE-STRENGTH TEST RESULTS

Screen/Slotted-Liner Construction	Screens/Liners Tested	Screen/Liner Size (in.)	ID (in.)	Average Collapse Pressure (psi)	Type Failure
Welded (1) screen	5	2⅞	2.441	5,600	Screen jacket against pipe base
Welded (1) screen	5	3½	2.992	4,400	Screen jacket against pipe base
Welded (2) screen	1	2⅞	2.992	10,000	No failure
Wire-wrapped screen (drilled pipe base)	1	2⅞	2.441	9,000	Screen and pipe base
Wire-wrapped screen (grooved and slotted pipe base)	1	2⅞	2.441	4,600	Screen and pipe base
Wire-wrapped screen (drilled pipe base)	1	3½	2.992	8,800	Screen and pipe base
Wire-wrapped screen (grooved and slotted pipe base)	1	3½	2.992	4,500	Screen and pipe base
Rod-based screen	1	2⅞	—	5,200	Screen
Rod-based screen	1	3½	—	4,500	Screen
Slotted liner	1	2⅞	2.441	9,200	Pipe
Slotted liner	1	3½	2.992	6,200	Pipe

because the coupling is the weakest link. Judging from the tensile strength test results, the failure loads tabulated in Table 5.1 are generally about 35% higher than the published minimum yield strengths of nonupset API tubing. To avoid screen failure, however, an equivalent load no greater than 55,000 psi should be placed on a screen when it is run in a well and set in position. Of all the pipe-based screens tested, no screen was clearly superior in terms of tensile strength.

Rod-based screens have sometimes been advocated for oilfield use. Here, the procedure has been to increase the number and diameter of longitudinal rods so that the screen's tensile strength and collapse resistance are enhanced. As the name implies, the screen is manufactured without an inner-drilled pipe base.

Tensile-strength tests identical to those described previously were conducted on the rod-based screens. Unlike the pipe-based screens, however, the failures in these screens occurred in the body (**Fig. 5.10**) or, more precisely, in the area where the screen section was welded to the pin connection. Apparently, the weakened condition existed there because of annealing during welding, lack of weld penetration, or lack of sufficient weld material. As Table 5.1 shows, the tensile strength of the rod-based screen is about half that of pipe-based screens.

Collapse-Strength Tests. Table 5.2 gives the results of the collapse tests. The failures noted in the Welded (1) screens were surprising: most occurred in the screen (jacket), leaving the pipe base unaffected. The failures here resulted from excessive standoff between the screen and the pipe base; once the failure strength of the screen was reached, it collapsed against the pipe base. When this occurred, a rather unusual herringbone pattern developed on the screen jacket. This pattern resulted because the longitudinal ribs were forced along the screen axis away from the pipe base (**Fig.**

Fig. 5.11—Collapse failure of Welded 1 screen.

Fig. 5.12—Collapse failure of Welded 1 screen.

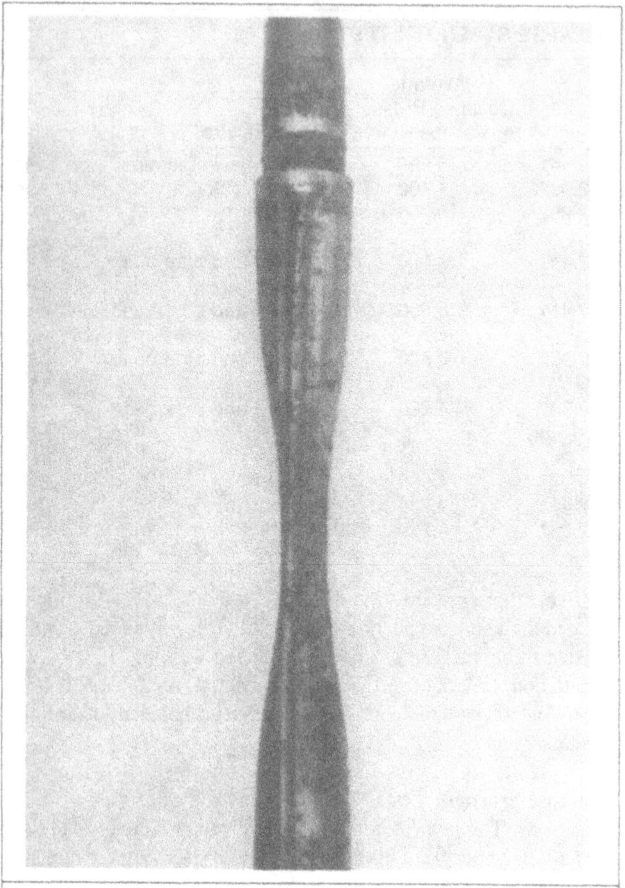

Fig. 5.13—Collapse failure of Welded 1 screen.

Fig. 5.14—Collapse failure of 2⅞-in. slotted liner.

5.11). It appeared that, the greater the standoff between the screen and the jacket, the lower the collapse resistance. In a number of cases, the collapse resistance was below the published values for Grade J-55 tubing, about 6,500 psi. When no standoff existed between the screen and pipe base, collapse strengths were much higher.

The second failure mode noted was similar to that discussed above, except that no herringbone pattern resulted. Instead, a portion of the screen was forced up in a welt (**Fig. 5.12**). The height of the welt was proportional to the standoff distance between the screen and the pipe base. Note in Fig. 5.12 that a welt is also present, indicating excessive standoff between the screen and pipe base. These types of failures have been experienced in the field as a consequence of screen plugging during gravel packing. In the third failure mode, the screen jacket and pipe base failed simultaneously. This occurred with some Welded (1) screens (**Fig. 5.13**).

Fig. 5.15—Cross section of a prepacked screen.

Rod-based screen failures resulted in a total collapse of the screen. The appearance of the rod-based screen after failure was similar to that shown in Fig. 5.13.

Collapse tests were also conducted on wire-wrapped screens. As mentioned previously, the screen is wrapped directly on the pipe base and is held in place by two longitudinal welds. The pipe bases of the screens were different: one had a drilled pipe base while the other was grooved and slotted.

The collapse-strength test results of these two screens were quite different. Those screens with drilled ⅜-in. holes were much stronger. Their collapse resistance exceeded 9,000 psi, compared with 4,500 psi for those with grooved and slotted pipe bases. The higher collapse resistance of the drilled pipe-based screens resulted because these screens allowed the applied forces to be distributed uniformly around the holes, thus increasing the load resistance. By contrast, the grooved and slotted pipe-based screens actually weakened the screen because the slots did not permit the applied load to be distributed uniformly. As a result, the collapse resistance was much lower.

The final collapse-strength tests were conducted on 2⅞- and 3½-in.-diameter samples of 0.020-in.-wide slotted tubing. Two tests were performed on the 2⅞-in. sample. In the first test, the sample did not fail at a pressure of 9,000 psi. Posttest examination of the sample showed that all the slots had been forced to close. In the second test, failure occurred at a pressure of 9,200 psi. Inspection of the failed test sample indicated that it appeared to implode (see **Fig. 5.14**). The 3½-in. sample failed at an applied pressure of 6,200 psi, which is about the published collapse pressure. Indications were that milled slots are not as resistant to compressive loading as round holes are.

Flow-Capacity Tests. Flow test results of all screens and slotted pipe under investigation are given in Chap. 3. Probably the most significant feature of the screens is that they have extremely high flow capacities. Even the slotted liners have flow capacities of about 1,000 B/D-ft with less than a 15-psi pressure drop across them.

However, wire-wrapped screens have significantly higher flow capacities and are not plugged as easily as slotted liners.

Prepacked Screens

Resin-coated gravel and wire-wrapped screen technology were combined to produce prepacked screens. These screens are fabricated by filling the annulus between two concentric wire screens with properly sized resin-coated gravel (**Fig. 5.15**). The resin usually is a phenolic, but other plastics can be used. The gravel pack is actually installed and consolidated at the surface and then is lowered into the well. Consequently, uncertainties in gravel placement can be avoided. On the other hand, because the prepacked screen must contain gravel, certain versions of the design have a substantial annular thickness that must be considered in terms of the casing size and the diameter of tools that can be run. Other designs merely fill the annulus between the wire wraps and the pipe base with the resin-coated gravel. As a result, they are the same size as standard screens but do not have a thick gravel layer.

While these devices have demonstrated excellent sand control capabilities, their ability to maintain sustained productivity when used as the sole means of controlling formation sand production has been disappointing, except in ideal situations. The problem appears to be that they are much more easily plugged with formation fines than wire-wrapped screens are because the prepacked screens have one less degree of freedom than wire-wrapped screens do from the standpoint of permitting particles to flow through their pore structure. A contributing factor may be that the formation sand migrates into the annulus around the screen, completely fills it, and also fills the perforation tunnels so that losses in productivity are inevitable. The same result has been suspected in wells where only a slotted liner or screen has been used to control formation sand. In this situation, however, plugging has not been observed to occur as rapidly as with prepacked screens. Prepacked screens are useful, as are slotted liners and screens, in controlling formation sand from coarse-grained reservoirs having a low percentage of fines and clay. Their use usually should be considered relatively short term, however, because of ultimate productivity impairment.

Gravel packing in the annulus between the prepacked screens has been used with success. However, gravel packing effectively requires that extremely clean equipment and fluids be used to avoid damaging the prepacked screen. Attention to cleanliness is much more critical in this situation because the prepacked screens are considerably better at filtering materials than screens are.

Prepacked screens sometimes are placed across the top of the completion interval opposite the upper perforations to prevent screen erosion if the gravel pack settles and exposes the bare screen. The idea here is that the resin-coated gravel will serve as a fluid diverter or blast joint and will avoid screen erosion in this location.

Other Considerations

Screens have corroded and/or cracked in certain situations. Corrosion has normally been detected on the pipe base of the screens, which is more corrosive than the stainless-steel jacket. Actually, a galvanic cell with a net potential of 0.3 V exists between the screen and the pipe base. Here, the pipe base acts as the anode and the stainless steel as the cathode. In highly corrosive situations, the flow of electrical current tends to increase the corrosion rate, but in normal service, no serious galvanic corrosion has been observed on pipe-based screens.

In certain high-temperature and high-salinity producing conditions, apparent stress cracking of the screen jackets has been observed when 304 stainless steel was used. To deter stress-corrosion cracking, the following screen materials may be considered as alternatives to 304 stainless steel.

Bottomhole Static Temperature (°F)	Screen Material
>180	Incoloy 825
150 to 180	316 stainless steel
150	304 stainless steel

Summary

Tensile Strength.

1. The primary tensile failure of pipe-based screens and slotted liners occurs in the 10-round API coupling. This failure results from yielding. The failure stress is normally about 75,000 psi for Grade J-55 tubing.
2. Rod-based screens usually fail at the point where the screen is welded to the pin connection and have about one-half the strength of pipe-based screens.
3. Pipe-based screens are about twice as strong as rod-based screens under tensile loading conditions.

Collapse Strength.

1. Collapse failures of Welded 1 screens relate primarily to the excessive standoff of the screen from the pipe base. If the standoff is low, the collapse resistance is high; if the standoff is high, the collapse resistance is low.
2. Welded 2 and wire-wrapped screens usually exhibit higher collapse resistance than Welded 1 screens do.
3. The collapse failures of wire-wrapped screens relate to the shape of the openings in the pipe base. Round holes support loads more effectively than slots do, creating a screen that has about twice as much collapse resistance.

Flow Capacity.

1. The flow capacities of Welded 1 and 2 and wire-wrapped screen available from the five manufacturers tested were comparable.
2. The flow capacities of screens, slotted liners, and gravel packs do not pose significant restrictions to flow unless they become plugged.
3. Rod-based screens offer no significant advantage over pipe-based screens in terms of flow capacity.

Prepacked Screens.

1. Prepacked screens provide excellent sand control but are easily plugged with formation materials. As a consequence, their completion lives may be limited owing to low well rates.
2. Gravel packing around the prepacked screens is an alternative to achieving sand control and sustained productivity, provided that the gravel-packing operation can be conducted without impairing productivity.
3. Prepacked screens have been used as the top joint opposite the upper perforations to help prevent screen erosion if the gravel pack settles and exposes the screen.

Reference

1. *Ground Water and Wells*, Johnson Div., UOP Inc., St. Paul, MN (1975).

SI Metric Conversion Factors

bbl	× 1.589 873	E−01	=	m^3
ft	× 3.048*	E−01	=	m
in.	× 2.54*	E+00	=	cm
lbf	× 4.448 222	E+00	=	N
psi	× 6.894 757	E+00	=	kPa

*Conversion factor is exact.

Chapter 6
Gravel-Pack Well Preparation

Introduction

The proper preparation of a well for gravel packing can be the key to completion success. Careful planning, well preparation, and completion execution are required to increase completion productivity and longevity. Most gravel-pack practices are interdependent, so it is important to execute all steps properly and in the correct sequence. If a well is prepared properly, it has the best chance of being placed on production undamaged and producing sand-free. To achieve these goals, attention must be given to drilling practices, cleanliness (casing, workstring, tanks, and completion fluids), completion fluids, filtration, perforating, perforation cleaning, acidizing, gravel specifications, gravel and equipment placement techniques, tool specification, and rig and service company personnel.

Drilling Practices

Drilling practices can affect a gravel pack the same way that they affect conventionally perforated wells. The well should be drilled to maintain borehole stability, and drilling fluids should be used that will not damage the formation to the point where the formation-damage depth is past the perforation penetration. Drilling-fluid filtrates should be compatible with completion fluids and should not interfere with completion operations. Ideally, the drilling fluid selected should be dense enough to result in a well that is slightly overbalanced, should have low fluid loss, and should be compatible with the clays in the productive formation.

Cleaning the Casing and Workstring

Cleanliness is one of the most important considerations in the implementation of gravel packs. The absence of any particulate materials can result in an undamaged completion. Particulates or other damaging material are usually, but not always, attributable to dirty workover fluids. Damaging material also can be on the inside or the outside of the workstring or inside the casing string. All particulates and other potentially damaging materials should be removed before the gravel pack is performed. This step can be costly and time-consuming, but should not be taken lightly. The cleanliness of the tubulars should never be ignored.

Casing. Potentially damaging materials in the casing include cement scale, rust, oxidized hydrocarbon deposits, and other debris. Removing these deposits requires bit and scraper trips for mechanical dislocation before they are circulated from the hole. The use of water, surfactants, hydrocarbon solvents, acids, and special particulate-suspending (flocculation) solutions may also be required to achieve clean casing.

Workstring. The workstring can also contain potentially damaging materials. Scale, rust, asphaltenes, and/or waxes may be present on the inner or outer surfaces. Both of these surfaces must be clean before the gravel pack is performed. These materials commonly are removed with solvents, either water or hydrocarbon-based, or weak acids, which react with nonsoluble scales or deposits. Sometimes the casing and tubing are scoured by circulation of a sand slurry through them at high rates. Once the workstring has been cleaned, every effort should be made to keep it clean.

During tripping out of the well before screen installation, attention to cleanliness is essential to avoid possible contamination. In particular, use of thread-lubricant compounds should be minimized because they can easily plug screens when mixed with solid particles.

Tanks and Lines. Tanks and lines should not be ignored. They should meet the same cleanliness requirements as other rig and/or well components. Such measures as scraping, jetting, and acidizing can be used for cleaning. Flocculating solutions have also been used effectively to remove sludge and particulates from lines and tanks. Where gravel packing is common, tanks should be dedicated to gravel-pack use to avoid repeated cleaning operations for drilling-mud removal.

Gravel-Pack Fluids

Gravel-pack fluids can be water- or oil-based. The water-based fluids are usually considered to be more flexible. Their densities, viscosities, and formation compatibilities are more easily controlled than those of oil-based fluids. Consequently, water-based fluids are used more commonly. In certain applications, however, produced crude has been widely used as the completion fluid. Its use usually depends on the cost and availability of alternative completion fluids. Crude oil is used when it is more attractive economically, provided that it exhibits the proper physical properties and can be handled safely.

The source of the completion fluid can vary. Regardless of its origin, the fluid should contain minimum particulate material and its chemistry must be compatible with the rock formation and connate water. Water that is too fresh may cause clays to swell or disperse, while the presence of some ions may cause precipitation when in contact with formation water.[1] The most common sources are field or produced brine, seawater, bay water, or fresh water.

The density of workover fluids should be controlled when possible with soluble salts such as sodium chloride, calcium chloride, calcium bromide, sodium bromide, zinc bromide, calcium nitrate, potassium chloride, and zinc chloride. All have particular advantages and disadvantages. Sometimes two or more of the brines are

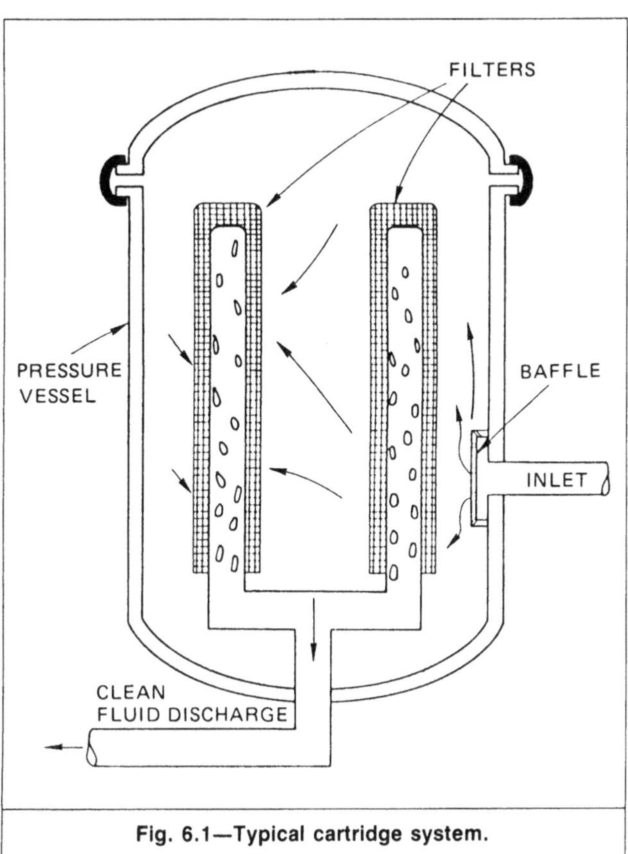

Fig. 6.1—Typical cartridge system.

mixed together to achieve the desired fluid properties. Of the fluids mentioned above, the sodium and calcium chlorides and the calcium bromide are most commonly mixed together and used for gravel packing. On the low-density side, foams have been used in certain low-bottomhole-pressure situations.

Viscosity is usually controlled with natural and synthetic polymers. Among them are guar gum and starch, XC-biopolymers (xanthan gum), and hydroxyethylcellulose (HEC), carboxymethyl cellulose and polyacrylamide. The amount of the polymers to be added to a brine depends on the viscosity desired, the type and concentration of the salts present, and the temperature. Because of its unique combination of low residue and good viscosity yield, HEC is a popular viscosity builder for gravel packing and other completion and workover applications.

Foams have also been used in certain gravel-pack situations. Foam is a simple mechanical mixture of air or gas dispersed in clean fresh water or field brine containing a small amount of surfactant. Surfactant type and concentration should be selected to develop a stable foam with the specific well fluids encountered. These fluids have been applied widely in shallow, low-pressure reservoirs. In wells with low fluid levels where circulation of oil- or water-based fluids is impossible, foams can be used for certain workover operations—i.e., washing out sand, drilling in, liner removal, tool recovery, and deepening.

Fluid-loss control is often necessary when wells are gravel-packed to prevent excessive loss of expensive or difficult-to-unload fluid or to prevent formation damage by particulate, hydrocolloidal viscosity binders, or ionic species present in the workover fluid. Fluid loss should be controlled by adjustments to the completion-fluid density when possible. When fluid-loss materials are required, they should be used so that they do not adversely affect the gravel pack.

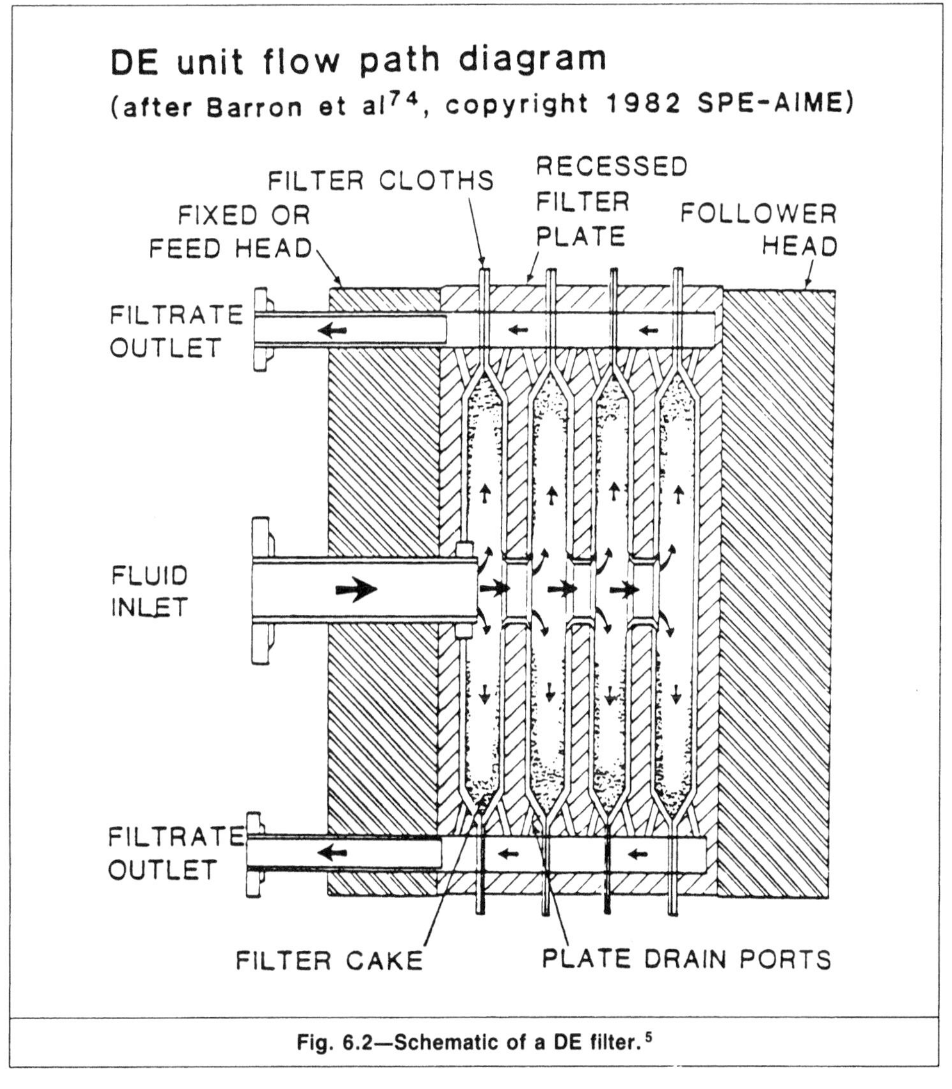

Fig. 6.2—Schematic of a DE filter.[5]

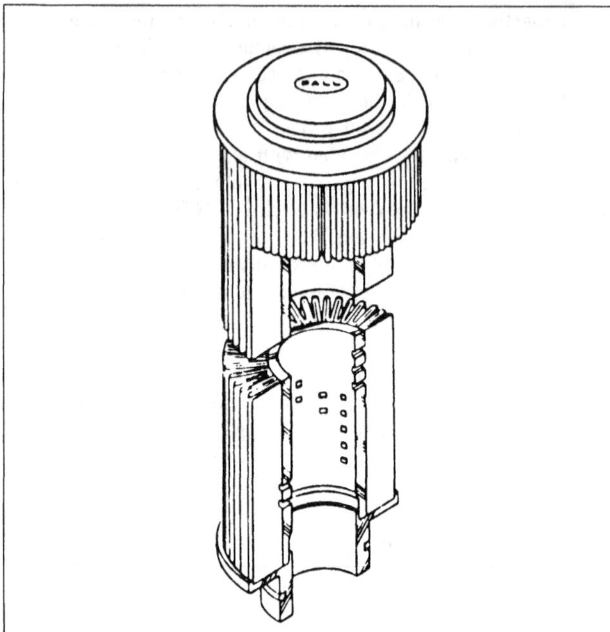

Fig. 6.3—Pleated filter element. (Courtesy Pall Well Technology.)

Ideally, fluid loss should be controlled at the formation face. This can be accomplished by incorporating into the workover fluid particulate materials of the proper size, shape, concentration, and particle-size distribution to form an impermeable filter cake without invading the formation. A filter cake should form quickly so that the movement of fluids and polymer in the pore space is minimized. The filter cake must be soluble or degradable in oil or acid. To meet this requirement, calcium carbonate, graded salt, lignosulfonate, and soluble resins have been used in the completion/workover fluids to control fluid loss.

Completion-Fluid Filtration

A gravel-pack completion fluid must be sufficiently clean that suspended particles do not plug or reduce the permeability of the gravel pack or the producing formation.[2,3] Filtration to 20 μm is recommended as a minimum requirement. In many situations, filtration to 1 μm is required. How the filtration is accomplished—by cartridge, diatomaceous[4] earth (DE), or other techniques—is not important. The cleanliness of the fluid is the key issue. **Figs. 6.1 and 6.2** show examples of a multielement cartridge filter and a DE filter unit. These units can provide clean fluids if proper procedures are followed. Current applications indicate that a completion fluid filtered by a DE filter or cartridge filter upstream in series with lower-rated cartridge filters probably is the cleanest fluid possible for many applications in terms of cost and time.[5-7] In many cases, however, only a cartridge filter is used when the completion/workover fluid is initially clean. The main advantage of the DE filter unit is its ability to filter dirty fluids at reasonably high rates compared with cartridge filter units.

The cartridge filter elements are usually made from string-wound polypropylene, graded-density polyester, or phenolic-impregnated cellulose polyester. These are rated from 1 to 100 μm or higher. The DE filter units continuously supply DE to a filter plate that acts as the filter medium to clean the fluid of impurities.

The number of cartridges required for a completion depends on the cleanliness of the workover fluid. Cartridge filter elements should be discarded once the pressure drop across them exceeds about 30 psi. Above this pressure level, the elements tend to break down and bypass or unload debris. The general cycle times for filters are based on the pressure drop that will permit them to perform as advertised.

The absolute-rated cartridge filters are a modification of the standard cartridge filter element but have pleated elements that offer about 15 times the surface area of conventional cartridges, as shown in **Fig. 6.3**. When in use, cartridge filters can be piped for either single- or multistage filtration (**Fig. 6.4**).

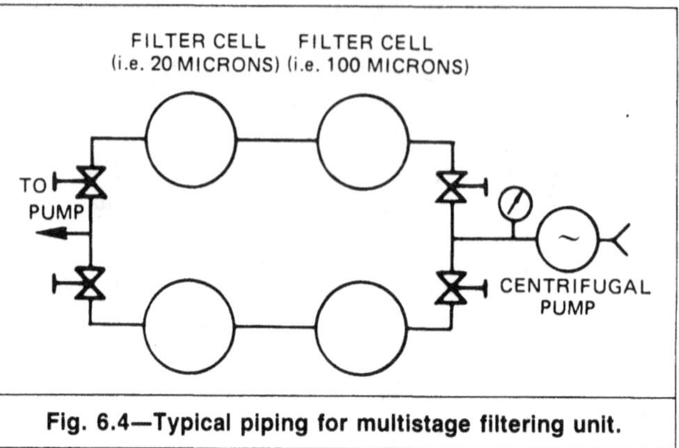

Fig. 6.4—Typical piping for multistage filtering unit.

Filtration should be conducted before viscosifiers are added to a workover fluid with polymers because at this point filtration tends to strip the polymers from the fluid and to plug the filters. To achieve the desired level of polymer mixing, fluids should be sheared in mixing devices that disperse the polymer uniformly in the completion fluid to eliminate unhydrated clusters, commonly called "fish eyes." Likewise, the filtration of naturally viscous fluids can be a problem because of rate and pressure-drop restrictions. In this case, the filter rating may have to be increased to achieve acceptable rates. If equipment, tanks, and time are available, a dirty completion fluid can stand overnight or longer to allow some of the particles to settle on the bottom of the tanks. Decanting the liquid from the tank allows the settled solids to be removed from the fluid.

Openhole Gravel Packs

Openhole gravel packing can result in a high-productivity completion but has the disadvantage that extraneous water and gas are difficult to isolate from the wellbore should they present a problem. Another potential problem with openhole gravel packs in the well preparation phase is the lack of hole stability. If hole stability cannot be maintained, a successful completion is doubtful. Even though openhole oil- and gas-well gravel packs are best suited for ideal reservoirs, they are often used in nonideal situations because of their high productivity when high water cuts are anticipated.

Once the decision has been made to use an openhole gravel pack in a well, the other decisions to be made relate to the casing point, total depth, and underreaming the completion interval.

Selecting the casing point is straightforward. It can usually be determined by running and correlating well logs or by analyzing sample cuttings at the surface. If at all possible, the casing point should always be selected so that the overlying shales are cased off because shales sometimes heave or slough and adversely affect a completion by contaminating the gravel pack.

Underreaming. Underreaming is recommended in most openhole completions. The advantage of underreaming is its provision for a thicker filtering medium between the screen and the formation. Because underreaming enlarges the hole diameter, it decreases the flow resistance into the well. Decreased resistance means higher well productivity compared with an identical well that has not been underreamed.

Underreamers, or hole enlargers, operate by the application of hydraulic pressure to the device, which is attached at the end of the drillstring. The pressure forces the arms out against the borehole as the drillstring is rotated. This enlarges the hole diameter to a predetermined radius. When the hydraulic pressure is released, the arms return to their original position, allowing the underreamer to be removed from the well. In the absence of perforations, hole enlargement by underreaming can remove wellbore damage and mudcake from the borehole. In some isolated situations, small-diameter screens are run in conventional wells. In this case, underreaming is probably not necessary because of the large annular volume around the screen.

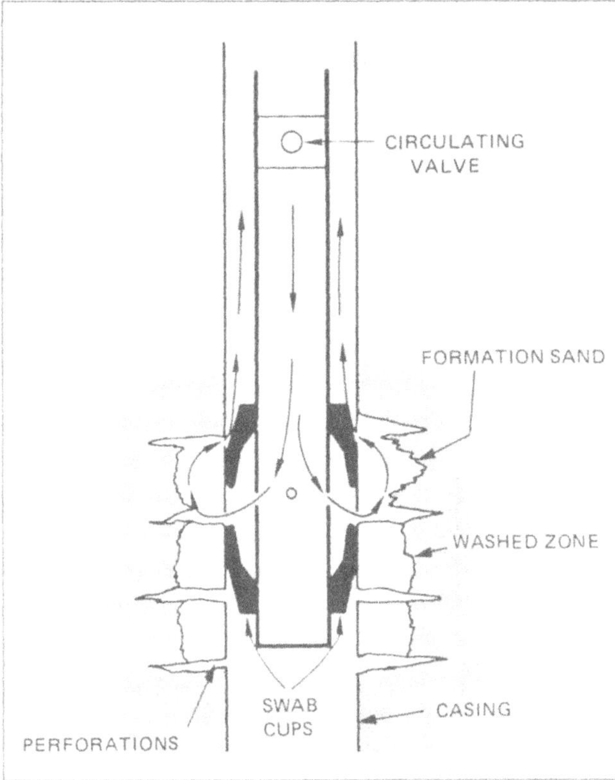

Fig. 6.5—Schematic of perforation wash tool, conventional circulation type.

The diameter of the underreamed interval usually depends on the diameter of the production casing and the sizes of the screen and slotted liner. The screen (slotted liner) should be sized to allow it to be washed over if it must be removed from the well. The underreamed section diameter should normally be no more than 18 in. in a well with 9⅝-in. casing and can be decreased for smaller casing and screen sizes. In addition to providing a larger gravel-pack thickness, the underreamed interval can also be drilled with a different, nondamaging fluid than that used to drill to the casing point. Formation damage is thereby minimized.

Underreaming fluids used for gravel packs should have a low solids content and should be nondamaging to the formation. The fluids of choice usually include polymers for viscosity and incorporate fluid-loss control. HEC and other polymers have been used in many cases for viscosity control because of their reasonably low residue, which tends to minimize potential formation damage compared with clay-based fluids. During underreaming, solids must continuously be removed from the underreaming fluid with desanders and/or desilters.

Where their use is feasible, brines make excellent underreaming fluids. Because of the low viscosity and potentially high fluid loss of brines, however, their capacity to clean and remove cuttings from the hole is limited unless high fluid velocities can be achieved in the well's annulus.

Calcium carbonate and lignosulfonate typically are added to underreaming fluids for fluid-loss control. Any fluid-loss-control material should be degradable or acid-soluble. Potassium chloride is used to prevent clay swelling. These fluids are available from almost all service companies that supply drilling or workover fluids. Some completion-fluid manufacturers have developed saturated-salt brines, which use salt particles to control fluid loss.

After the completion interval is underreamed, caliper logs should be run over the interval to provide a basis for future gravel-volume requirements and to examine the effects of drilling rate, pump rate, fluid loss, and other variables on the openhole variable. Polymer pills are commonly used to maintain wellbore stability and fluid-loss control after underreaming until the gravel-pack equipment is placed and the well is subsequently gravel-packed.

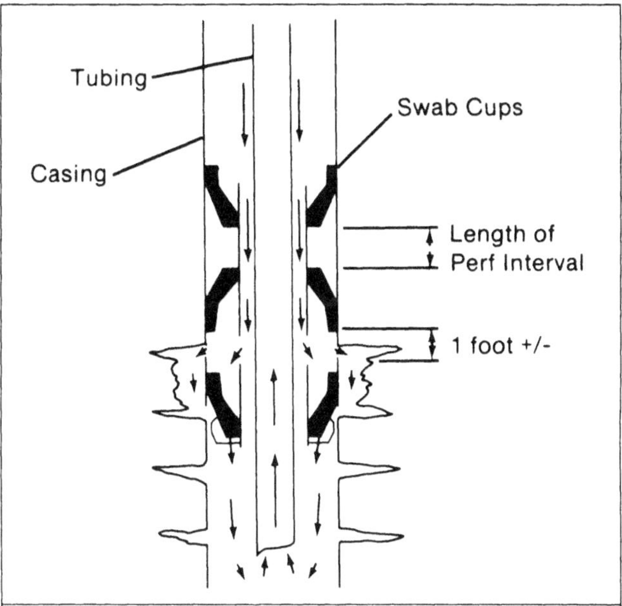

Fig. 6.6—Schematic of perforation wash tool, reverse-circulation type.

Summary of Recommendations for Completing Openhole Gravel Packs

1. If possible, select a casing point to case off all overlying shales.
2. Underream the completion interval to the largest practical hole diameter.
3. Underream with a low-solids, acid-soluble, or degradable fluid-loss material.
4. Clean the casing, tanks, and all lines before running the gravel-pack equipment.
5. When running the screen in the well, have tools in a position so that there is no flow through the screen. This minimizes screen plugging.
6. Gravel pack the well using appropriate fluids, equipment, and procedures.

Cased-Hole Gravel Packs

Cased-hole gravel packs differ from openhole gravel packs in that the well screen is run inside perforated casing (sometimes referred to as inside gravel packs). The primary difficulty with this type of completion is its potential inhibition of well productivity of high-rate wells because the gravel-filled perforations are the main restriction to flow. The design of cased-hole gravel packs should be based on fluid inflow into the well. Consequently, selecting the appropriate size and number of perforations and effectively prepacking them with gravel are critical for high-rate wells. Well cleanliness is extremely important because damage can also severely limit well productivity.

The selection of perforating equipment and procedures deserves great attention before cased-hole gravel packing, not only on a new well but also on wells being worked over. Many wells are not initially gravel-packed. They may produce well, but excessive sand production may subsequently require sand control. Gravel packing at this point without reperforating could result in a low-productivity well if the perforation area is inadequate for a gravel pack.

Perforating Practices. The charges of some gravel-pack perforating guns can create perforations with diameters well over 0.75 in. Usually, 5- or 7-in. perforating guns shoot the large holes. Should large-diameter perforations be considered necessary for a particular completion, larger casing strings than originally recommended may be necessary.

Once the perforation size and density have been determined, perforation phasing should be considered. Phasing can be 0, 90, 120, or 180°; however, most 3⅛-in. and larger guns are phased at 90

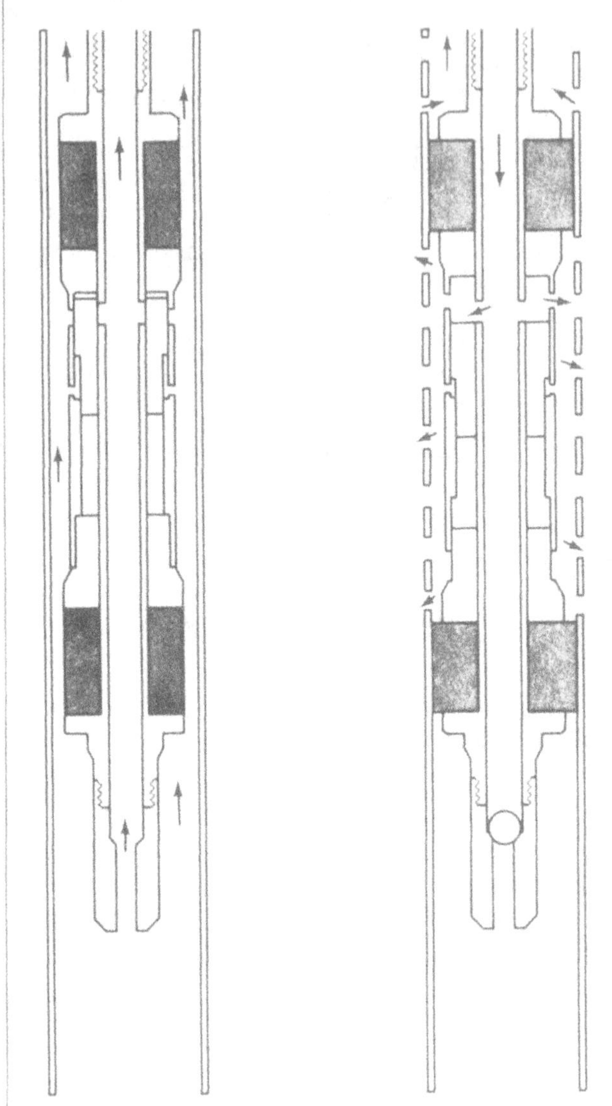

Fig. 6.7—Schematic of perforation wash tool, inflatable-packer type. (Courtesy B-J Hughes Services.)

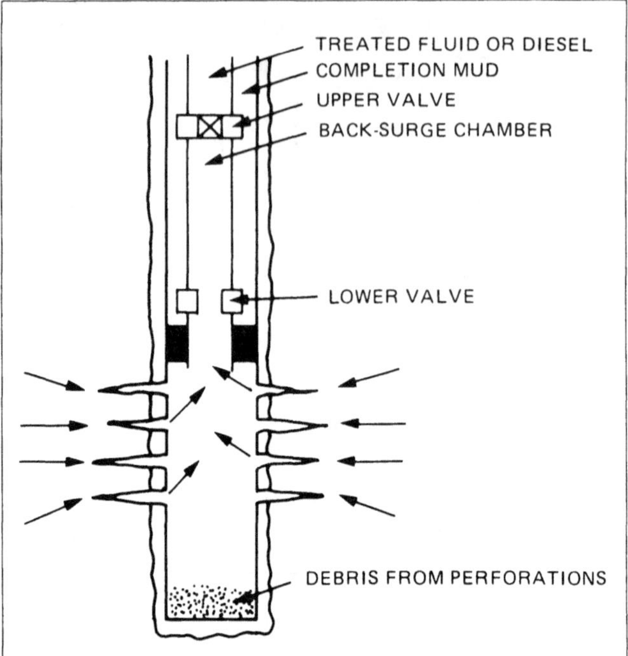

Fig. 6.8—Schematic of backsurging tool illustrating principle of the technique (after Suman *et al.*[8]).

or 120°. There are some indications that 90° phasing is the most effective.

The perforation size as it relates to gravel packing is most important. The size controls not only the amount of formation sand exposed to the well but also how effectively gravel-pack material can be placed through the perforations. In both cases, large-diameter perforations are desirable because the perforation area through the casing and cement sheath is more important than penetration as long as the formation is in communication with the well. As a general guideline, select perforating guns that yield the largest-diameter perforations yet allow adequate communication between the well and the producing formation.

The perforation density necessary to produce a well at the desired rate is also critical. Perforation densities are typically between 8 and 12 shots/ft for cased-hole gravel packs. The nodal analysis techniques discussed in Chap. 3 can be used as a guideline in selecting both the number and size of the perforations.

Underbalanced perforating is the recommended practice.[8-12] How much underbalance is necessary depends on the formation and the fluid properties. One general guideline is to perforate oil wells 500 psi and gas wells 1,000 psi underbalanced, although specific situations may dictate different procedures. The underbalance pressure should not exceed the collapse strength of the casing or cause excessive formation failure that prevents the removal of the guns from the well. Some operators[11] advocate flowing the well for several tubing volumes to facilitate perforation cleaning.

Both wireline and tubing-conveyed perforating practices are amenable to gravel-packed completions. While both have advantages and disadvantages, the tubing-conveyed guns are attractive in that they can perforate an entire interval underbalanced. In long intervals, the tubing-conveyed guns may also be less expensive than wireline guns. Regardless of which perforating technique is used, a cased-hole gravel pack should not be completion-limited by insufficient perforation area. Each gravel pack should be perforated from the standpoint of the production rate anticipated.

Perforation Cleaning. Perforation cleaning is critical to gravel-packed completions to ensure that formation damage is minimized and that the condition of the perforation tunnels and the region outside the perforations will allow gravel packing. Chap. 3 discusses the effect of packing the perforation tunnels with gravel on productivity. Methods for cleaning perforations include perforating washing, surging, underbalanced perforating, and in some cases producing the well for a period of time before gravel packing. These techniques should remove formation sand from all perforations so that high-permeability gravel can be pumped into them to maximize productivity and completion longevity.

Perforation washing[13-16] is performed by pumping a wash fluid through the perforations over a short interval (usually about 1 ft or less) to remove formation sand from the perforations and possibly from the region outside the perforations before prepacking with gravel. The amount of material removed varies, depending on the strength of the formation. More formation sand can be removed by washing in low-compressive-strength formations than in high-compressive-strength areas. How much formation material is removed can also be related to perforation wash techniques and procedures.

Fig. 6.5 shows a conventional circulation-type perforation wash tool. Several different designs allow normal and reverse circulation washing. Some tools have opposed swab cups and inflatable elements. **Fig. 6.6** illustrates a reverse-circulation tool; **Fig. 6.7** shows an inflatable-packer-type tool.

Perforation washing is an art that requires attention to detail. The following procedures should be followed to achieve effective perforation cleaning and well control during washing operations.

1. Run and retrieve swab-cup perforation wash tools slowly. Running or retrieving them at rates exceeding about 1,000 ft/hr risks either breaking the formation down or swabbing the well in.

2. Check the fluid level in the workstring every few stands when running or retrieving operations are in progress.

3. Select a wash-fluid density considered to be the minimum for well control.

4. Use a clear brine as the wash fluid. Do not use viscosity-building additives because they create filtercake buildup and reduce formation sand removal.

5. During washing, pump at the maximum practical pump rate and/or surface pressure without breaking the formation down. In most cases, 2 to 3 bbl/min is an *insufficient* pump rate to clean the perforations effectively.

6. Space the wash cups (elements) as close as possible (6 in.).

7. Wash from the bottom perforated interval up.

8. Use a wash volume of 10 bbl or until a pressure break is observed.

Clean fluids that are compatible with the formation should always be used when perforations are washed. Also, the density of the wash fluid should correspond to the minimum acceptable overbalance pressure. Both conditions are important to minimize the risk of damaging the formation with particulates or workover fluids. The loss of workover fluids during perforation washing operations can be a serious problem. When loss of workover fluids cannot be controlled, perforation cleaning with surge tools may be a viable alternative.

Perforation surging is another way to clean perforations and remove plugging material. The tool is no more than a cylindrical chamber that can be sealed at the surface so that the air inside the chamber is at atmospheric pressure. **Fig. 6.8** is a schematic of the tool. After the tool is run in the well, it is opened to expose the perforations to atmospheric pressure. The pressure change forces a limited volume of perforation debris, mud, formation sand, and reservoir fluid from the perforation. Compared with a wash tool, the surge tool requires tripping the work string for each surge run. Because flow through the tool is limited when the tool is opened, several trips may be required for long intervals. In some situations, the surge volume in the tool can be changed to create larger surge volumes.

Tests[14] have shown that surging does not remove formation sand as uniformly as perforation washing, but wells that have been surged show excellent productivity. When perforation washing is not used, surging may be a viable alternative to cleaning perforations.

Summary of Recommendations for Completing Cased-Hole Gravel Packs

1. As a minimum cleaning technique, the casing should be scoured with a bit and scraper. The casing fluid is then flushed until returns are clean.

2. The perforating equipment used should be the best compromise between perforation diameter and penetration. Diameter is more critical than penetration, however, as long as there is communication with the formation. The perforation inlet area should not significantly limit well productivity. Because about two-thirds of the perforation area is occupied with gravel, a well perforated with 12 shots/ft has the effective flow area of an area perforated at about 4 shots/ft. Hence, guns with high shot densities are preferred for gravel-packed completions.

3. The well should be perforated underbalanced.

4. Perforations should be washed or surged to ensure that they are open and in a condition to accept gravel. Flowing the well is another option to perforation cleaning.

5. The screen/liner should be run in a clean well fluid. There should be no flow through the screen/liner during running in the hole to avoid plugging.

6. Compatible, clean, filtered fluids should be used at all times.

Acidizing

Acidizing can be used several different ways to enhance the success of a gravel-pack completion. Acid treatments can be performed either before or after the gravel pack using the design shown in Chap. 12.

When acidizing is performed before the gravel pack, the treatment is run as it would be in a normal well. If acidizing is considered necessary to completion success, it should be performed before gravel packing.

A well is typically treated with acid after the gravel pack if it is not producing at an acceptable rate. These treatments should be pumped at rates and pressures that will ensure that the formation will not be broken down.

References

1. Scheuerman, R.F. and Bergersen, B.M.: "Injection-Water Salinity, Formation Pretreatment, and Well-Operations Fluid-Selection Guidelines," *JPT* (June 1990) 836–45.
2. Sparlin, D.D. and Guidry, J.P.: "Study of Filters Used for Filtering Workover Fluids," paper SPE 7005 presented at the 1978 SPE Formation Damage Symposium, Lafayette, LA, Feb. 15–16.
3. Olivier, D.A.: "Improved Completion Practices Yield High Productivity Wells," *Pet. Eng.* (April 1981).
4. Barron, W.C., Young, J.A., and Munson, R.E.: "New Concept—High Density Brine Filtration Utilizing a Diatomaceous Earth Filtration System," paper SPE 10648 presented at the 1982 SPE Formation Damage Symposium, Lafayette, LA, March 24–25.
5. Jeffries, R.G.: "New Development in the Filtering of Solids-Free Completion Fluids," paper presented at the 1982 API Pacific Coast Meeting, Ventura, CA, Oct. 12–14.
6. Sharp, K.W. and Allen, B.T.: "Filtration of Oil Field Brines—A Conceptual Overview," paper SPE 10657 presented at the 1982 SPE Formation Damage Symposium, Lafayette, LA, March 24–25.
7. Suman, G.O. Jr., Ellis, R.C., and Snyder, R.E.: *Sand Control Handbook*, second edition (1983).
8. Bell, W.T.: "Perforating Techniques for Maximizing Well Productivity," paper SPE 10033 presented at the 1982 SPE Petroleum Exhibit and Technical Symposium, Beijing, March 19–22.
9. McLeod, H.O. Jr.: "The Effect of Perforating Conditions on Well Performance," *JPT* (Jan. 1983) 31–39.
10. King, G.E., Anderson, A.R., and Bingham, M.: "A Field Study of Underbalance Pressure Necessary To Obtain Clean Perforations Using Tubing-Conveyed Perforating," *JPT* (June 1986) 662–64; *Trans.*, AIME, **281.**
11. Ledlow, L.B., Sauer, C.W., and Till, M.V.: "Recent Design, Placement, and Evaluation Techniques Lead to Improved Gravel Pack Performance," paper SPE 14162 presented at the 1985 SPE Annual Technical Conference and Exhibition, Las Vegas, Sept. 22–25.
12. Bruist, E.H.: "Better Performance of Gulf Coast Wells," paper SPE 4777 presented at the 1974 SPE Formation Damage Symposium, New Orleans, Feb. 7–8.
13. Tausch, G.H. and Corley, C.B. Jr.: "Sand Exclusion in Oil and Gas Wells," *Drill. & Prod. Prac.*, API, Dallas (1959).
14. Penberthy, W.L. Jr.: "Gravel Placement Through Perforations and Perforation Cleaning for Gravel Packing," *JPT* (Feb. 1988) 229–36; *Trans.*, AIME, **285.**
15. Bonomo, J.M. and Young, W.S.: "Analysis and Evaluation of Perforating and Perforation Cleanup Methods," *JPT* (March 1985) 505–10.
16. Bruist, E.H., Jeffries, R.G., and Botts, T.M.: "Well Completions in the Beta Field Offshore California," paper SPE 11696 presented at the 1983 SPE California Regional Meeting, Ventura, March 23–25.

SI Metric Conversion Factors

bbl	× 1.589 873	E−01	=	m^3
ft	× 3.048*	E−01	=	m
in.	× 2.54*	E+00	=	cm
psi	× 6.894 757	E+00	=	kPa

*Conversion factor is exact.

Chapter 7
Packing the Perforation Tunnels

Introduction

Prepacking perforations in cased-hole gravel packs involves placing gravel through the perforation tunnels and in the cavity at the entrance of the perforation that was created by perforation washing, surging, or prior sand production. This practice is essential for long-life, high-productivity, cased-hole gravel packs,[1] as discussed in Chap. 3. Field results verify the importance of this practice (see Chap. 9). During prepacking, gravel is pumped in a transport fluid that must subsequently be dehydrated after gravel is placed in the perforations. "Dehydrated" means that there is grain-to-grain gravel contact and that the gravel is no longer pumpable. The ideal prepack gravel volume pumped through any perforation is equal to the perforation tunnel volume plus the volume of voids outside the perforation. This volume is normally a fraction of a cubic foot per net foot of perforation for new completions, but it may be substantially higher in wells that have previously produced formation sand. **Fig. 7.1** is a schematic of a procedure used to place gravel through the perforations. An alternative is to prepack and gravel pack the well simultaneously with the screen in place. This procedure is common in slurry-pack gravel placement.

The purpose of the prepack is to filter formation fines outside the perforation so that they do not enter it and thereby restrict productivity. The prepack prevents perforation plugging provided that the proper gravel size has been selected. The same method used to design the gravel pack should be used to select the prepack gravel size. Also, proper prepack placement techniques are required so that productivity can be maintained throughout the life of the well.

Gravel Transport Through Perforations

Achieving an effective prepack involves mixing the gravel at the surface with a carrier fluid, transporting it in the well to the perforated interval, and then placing it in the perforation tunnels. How effectively the gravel is placed into the perforations is a function of the pump rate, the fluid viscosity, and the size of the particles being transported.[2,3] For the gravel to be transported from the casing into the entrance of the perforations, the viscous forces on the particles must exceed the gravitational and inertial forces on the remaining particles. The particles that do not enter a particular perforation either are transported into lower perforations or are concentrated lower in the well. The factors governing whether a particle will be placed in a perforation on its downward descent include flow rate through the perforation, gravel size, fluid viscosity, and location of the particle relative to a perforation. The perforations, however, will eventually fill when the casing is filled with gravel above the perforations, provided that some fluid is lost and there are voids to accept the gravel. **Fig. 7.2,** a schematic of the prepacking process, illustrates some of these factors. A critical condition for effective prepacking is that the perforations must accept fluid. Low or no flow through a perforation results in packing only the entrance to the perforation (**Fig. 7.3**). This condition leads to low productivity and poor completion success because formation sand can migrate into the perforation tunnel. To enter the perforation, however, the gravel must be compatible with the perforation diameter, particularly if the gravel concentration is high; otherwise, bridging on the outside of the perforation may occur (**Fig. 7.4**).

Once a particle enters the perforation, it can be packed effectively regardless of whether low- or high-viscosity fluids are used, as long as fluid is lost into that perforation. The primary benefit of the high-viscosity fluid is that it assists the particle in entering the perforation on its downward descent. **Fig. 7.5** shows gravel-transport efficiency into the entrance of the perforation. How the gravel is transported into the perforation once it arrives at the entrance depends on the viscosity of the transport fluid. If the transport fluid is water, the gravel is transported over the top of a dune that forms initially at the entrance of the perforation (**Fig. 7.6**). When the dune reaches the end of the perforation, the channel over the top of the dune is the last to be packed. This depositional sequence is self-limiting in that low perforation flow rates will completely deposit the gravel in a horizontal perforation. Because water has low viscosity, no node development on the perforation occurs.

Fluid loss through the perforations is also required if viscous fluids are used to pack the perforations. Gravel is transported during perforation packing with a viscous fluid, as illustrated schematically in **Fig. 7.7,** as long as viscosity and flow rate are sufficient to transport the gravel to the end of the perforation. If the fluid viscosity has broken or flow rates are low, the packing conditions shown in Fig. 7.6 will occur. For the viscous-fluid, high-flow-rate case in Fig. 7.7, however, the gravel is initially transported and dehydrated at the end of the perforation and dehydration proceeds toward the entrance of the perforation. Because viscous forces dominate in this situation, gravel nodes form at the entrance to the perforations as a result of viscous-drag effects. Packing each perforation actually helps divert fluid from the packed perforation into the remaining perforations, regardless of whether low- or high-viscosity fluids are used.

The effect of flow rate, particle size, particle density, and fluid viscosity on transporting a particle into the entrance of a perforation is further illustrated in **Fig. 7.8.** Note the effect of fluid viscosity

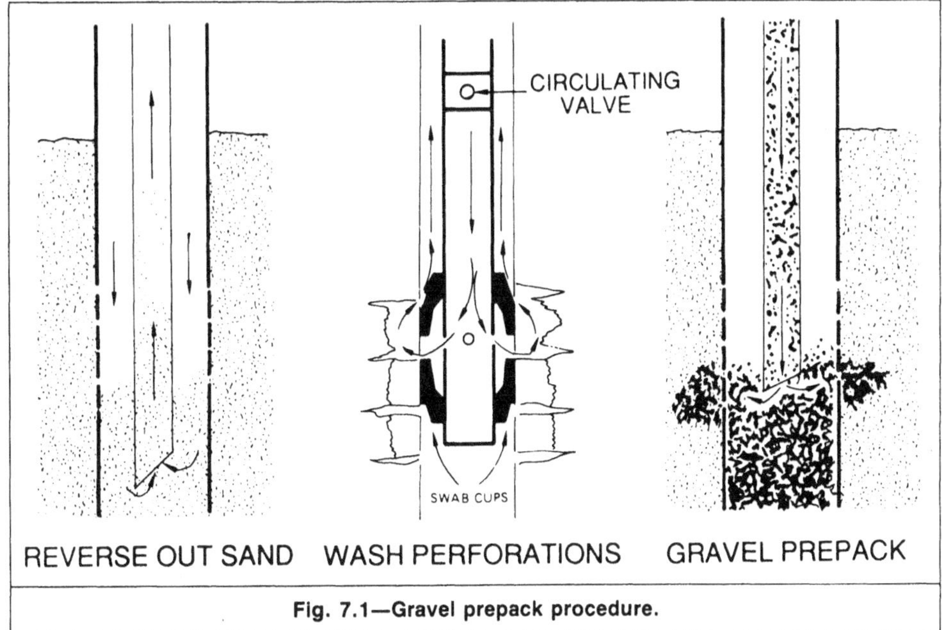

Fig. 7.1—Gravel prepack procedure.

and particle size. Fortunately, during packing, each perforation has several chances to be packed because numerous particles pass each perforation during pumping and because the rathole is continually being filled and thus higher gravel concentrations are lower in the well. Also, at some point, the settled-gravel interface will rise in the well so that it passes each perforation. This effect is also shown in Fig. 7.8—the lower perforations are accepting higher concentrations than the upper perforations are. Note that the viscous carrier fluid leads to a more uniform distribution of the particle concentration into the perforations caused by increased drag forces. This increases the transport efficiency of the fluid. For these reasons, viscous fluids usually have been preferred for perforation packing.

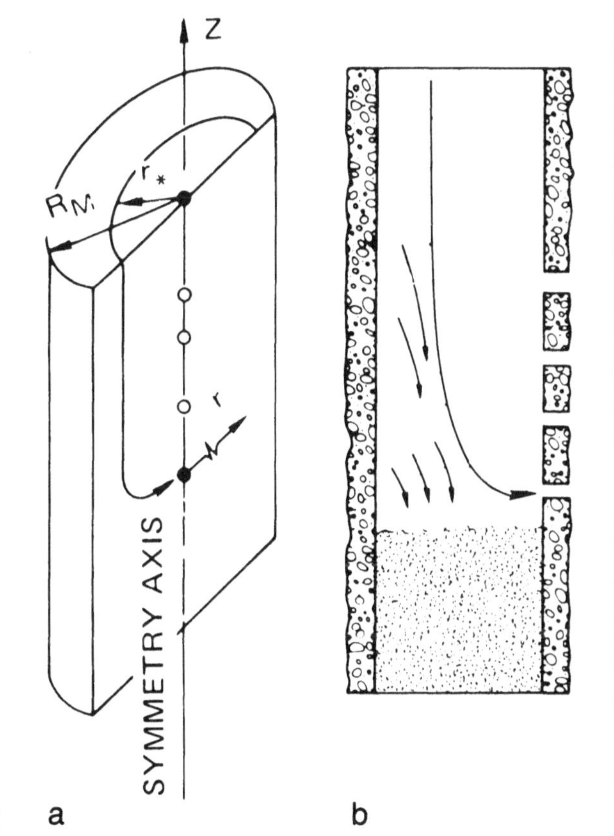

Fig. 7.2—Schematic of gravel prepacking efficiency: (a) mathematical model; (b) perforated casing.[3]

Field pump rates usually vary from <1 bbl/min to >5 bbl/min during perforation packing operations and are dependent on fluid viscosity and surface pressure. Pump rates usually should be on the high side to ensure placement in the perforations and to dehydrate the carrier fluid. The treating pressure should be below that required to break down the formation.

Uniform particle distribution is believed to be more desirable than high-density slugs of gravel because it prevents concentration buildup in the lower portion of the well. Also, particle buildup in the bottom of the well could result in incomplete packing if bridging occurs in the entrance to the perforations. Should perforations low in the completion interval not accept fluid, a gravel bank will progressively rise in the well to the lowest perforation that will accept fluid. Once this occurs, gravel placement into the perforations at this level in the completion will progress until the perforations are filled. The bank height will increase and placement will be into higher perforations, and so on.

Wellbore Angle Effects

Wellbore angle may affect the transport of gravel through high-

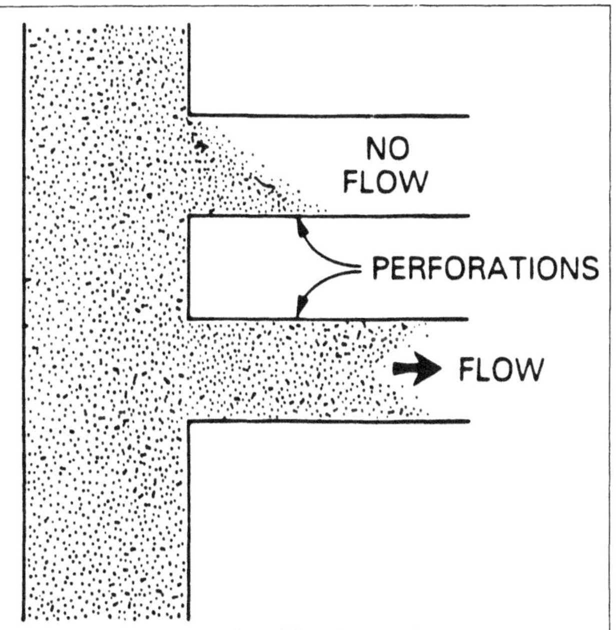

Fig. 7.3—Gravel placement in perforations with and without leakoff.

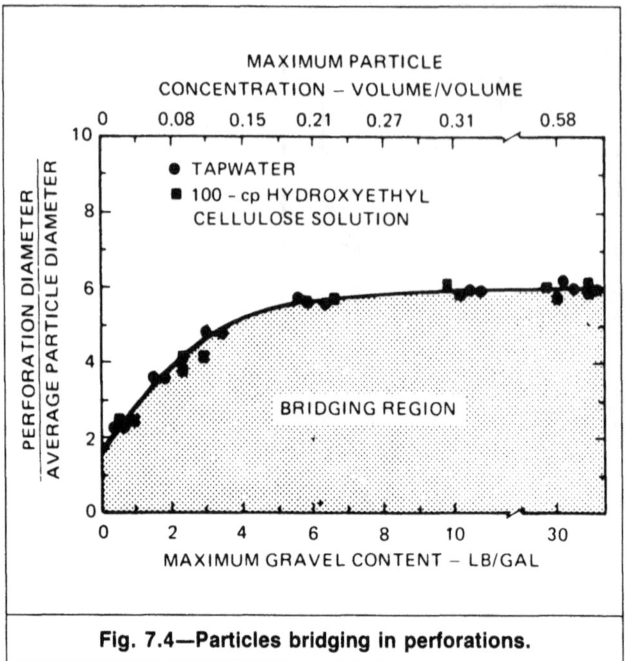

Fig. 7.4—Particles bridging in perforations.

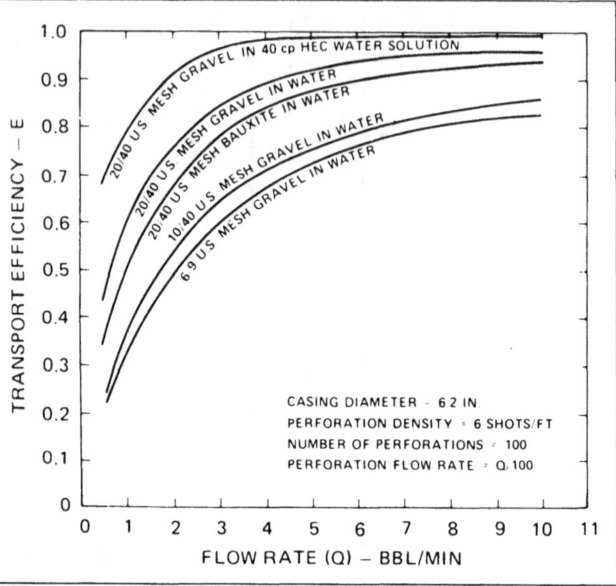

Fig. 7.5—Effect of flow rate on transport efficiency (calculated).

side perforations. To place gravel in perforations on the high side of the well requires sufficient flow to exceed gravitational forces. In very-high-angle wells, the gravel slurry may be distributed unevenly over the casing cross section. Experiments have shown that under these conditions the particles separate from the fluid and accumulate on the low side of the casing (see **Fig. 7.9**). In this event, the prepacking of low-side perforations is excellent but the prepacking of high-side perforations may be poor because of low perforation flow rates and lower gravel concentrations. Under these prepacking conditions, fluid transportation continues until all the perforations on the lower side of the casing are filled with gravel. At this point, the gravel slurry fills the rathole to the lowest high-side perforation. If the transport fluid exerts sufficient drag forces on the gravel particles, the gravel will then be transported into the high-side perforations, starting at the bottom of the wellbore and proceeding upward.

From a practical standpoint, prepacking the high-side perforations in a deviated well is possible with either water or high-viscosity fluids. Model tests have confirmed this.* However, the problem with prepacking before screen installation is keeping the gravel in these perforations during tripping out of the well. Two solutions are to perforate the well only on the low side and to prepack and gravel pack simultaneously with the screen in place.

*Unpublished report, Exxon Production Research Co., Houston.

Gravel Geometry Outside the Perforation

Research has shown that the gravel geometry outside the perforation in unconsolidated sands depends on the conditions that exist in the reservoir sand around the perforation.[4] When a gravel slurry is pumped through perforations where there has been no prior production, packing outside the perforation will occur only if the formation breakdown pressure is exceeded. In this event subsequent pumping results in pressure-parting the unconsolidated sand around the well. The parting is planar, gravel/formation-sand interface is sharp, and little mixing of the gravel and formation sand occurs. The orientation of the pressure part is normal to the least principal stress.

If sand production has occurred before gravel packing, the gravel geometry is the shape of the void that existed opposite the perforation. In many cases, however, a low-stress-state sand occupies this region and is pressure-parted. The orientations of these pressure partings are also normal to the least principal stress.

The gravel geometry after perforation washing is the shape of the void created by washing. If a low-stress-state sand exists here, it will also be pressure parted by the fluid if minimum pressure requirements are exceeded.

In the event that a formation is underbalance-perforated or surged, the shape of subsequent gravel placement outside the perforation tends to be erratic. The geometry tends to be controlled by the vari-

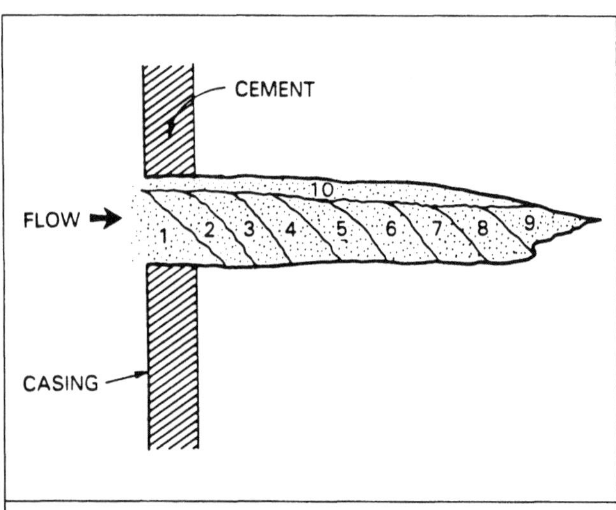

Fig. 7.6—Schematic of the perforation packing process (water as placement fluid).

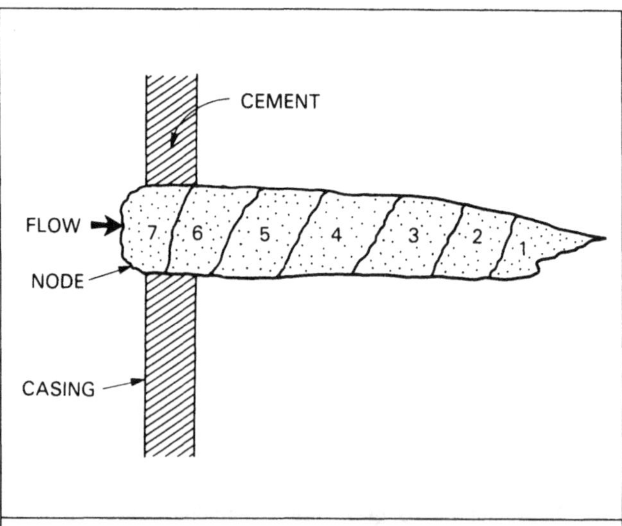

Fig. 7.7—Perforation-packing sequence with viscous fluids.

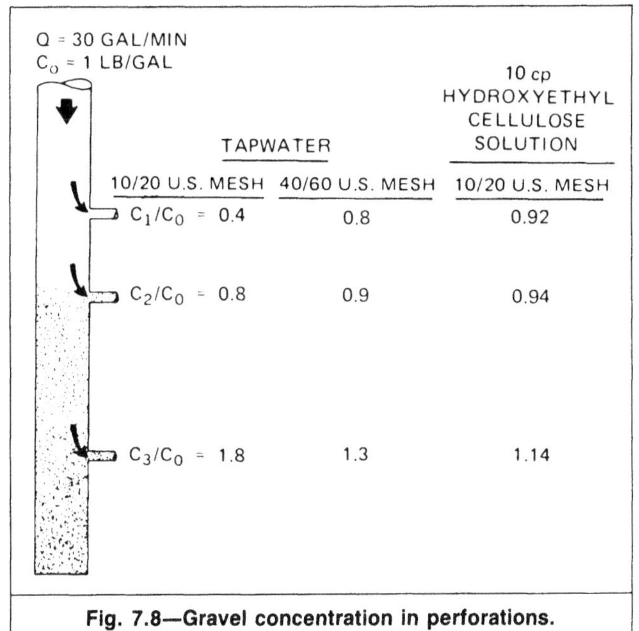

Fig. 7.8—Gravel concentration in perforations.

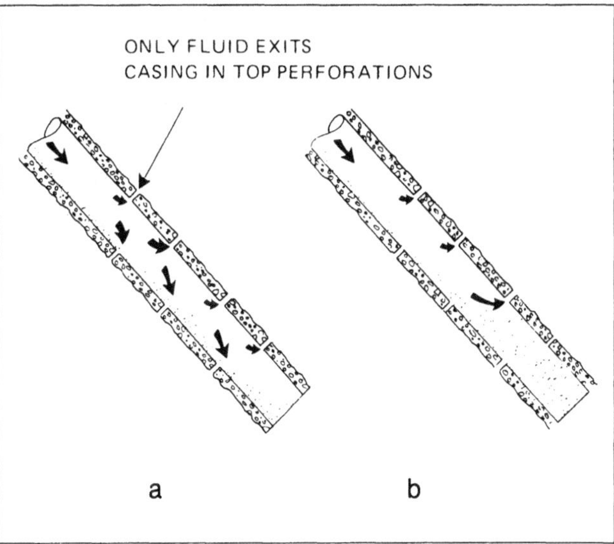

Fig. 7.9—Gravel prepacking in deviated wells: (a) gravel particles transported into the bottom perforations; (b) upper perforations filled.

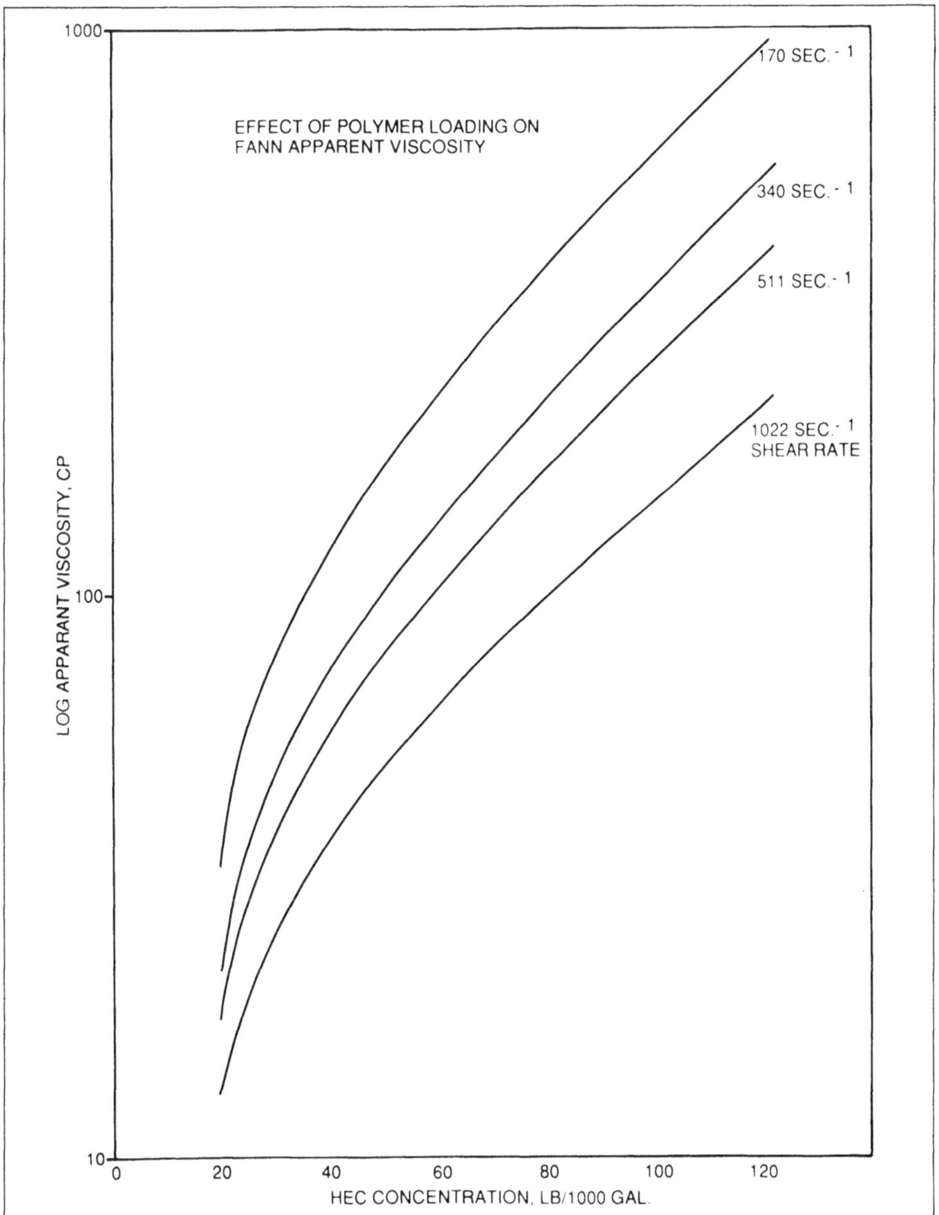

Fig. 7.10—Effect of polymer loading on Fann apparent viscosity.

able stress state of the formation sand over the completion interval. As a result, considerable intermixing between gravel and formation sand has been observed. High-productivity gravel packs are possible, however, in spite of the intermixing as long as the gravel has continuity and communicates with the perforations.

The consensus is that gravel usually should be placed through the perforations when pumping is conducted under matrix conditions and the formation is not broken down. The benefits of breaking the formation down are debatable despite the fact that some operators have done so. While this practice has been successful in some situations, breaking down the formation during the prepacking operation is not the usual approach.

Cleaning the perforations is believed to be key to effective gravel placement in the perforations. The most desirable condition is having a void outside the perforation that allows the gravel to be placed in this cavity. If no void exists, however, diversion techniques using staged treatments consisting of acid followed by gravel transported in hydroxyethylcellulose (HEC) fluids have been developed. The methods are said to enhance the amount of gravel placed compared with conventional circulation packing procedures.[5]

Guidelines for Fluid Viscosities and Gravel Concentration

Previous discussions showed that fluid viscosity has a significant effect on the transport of particles into perforation entrances. The high-viscosity fluids also transport particles in tubulars more effectively, particularly in deviated wells. Gravel-transport fluids are usually viscosified with polymers, such as HEC. Typical HEC polymer loadings or concentrations are 10, 20, 30, 60, or 100 lbm/1,000 gal or more. **Fig. 7.10** shows the associated average viscosities of these HEC polymer concentrations.

HEC polymer solutions are shear-thinning (power-law type) fluids whose viscosities decrease with shear rate. Depending on well conditions, the shear rate at a number of locations will change throughout the pumping sequence, and as a result, the viscosity changes. For this reason, Fig. 7.10 lists typical viscosities over a range of shear rates. Most perforation and gravel-packing operations are subject to periodic cessation of pumping. During these periods, the gravel will settle. While a certain amount of settling can be tolerated, gross settling before placement can create major problems with job execution. Hence, in the selection of the appropriate viscosity for a gravel-pack fluid, the possibility of pumping shutdowns should be anticipated.

To account for settling problems, fluids with slightly higher viscosities should be selected than would otherwise be used where no shutdowns occur. **Table 7.1** lists the gravel concentrations usually associated with various polymer loadings. Although these concentrations are representative, particular treatments may dictate deviation from these concentrations. Viscosity breakers should be added to polymer-laden fluids to reduce the viscosity substantially before the well is placed on production. Viscous fluids should always be sheared and filtered before they are pumped into the well.

TABLE 7.1—AVERAGE GRAVEL CONCENTRATIONS IN WATER-BASED POLYMER FLUIDS

	Polymer Loading (lbm/1,000 gal)						
	0	20	40	60	80	100	120
20/40 U.S.-mesh gravel concentration*	0.5	1	4	10	12	14	16

*lbm gravel/gal of fluid—i.e., the total volume is greater than 1 gal.

Summary

1. Packing the perforations significantly affects well productivity and completion life (see Chaps. 3 and 9).
2. Effective perforation packing requires that the perforations be sufficiently large that gravel bridging will not occur.
3. The proper size of gravel should be selected. The same gravel size should be used in the prepack and the gravel pack.
4. The perforations should be clean and capable of accepting injection fluid through the perforations before packing; i.e., there should be fluid loss.
5. The perforations can be effectively packed with either water or viscosified fluids. Viscous fluids provide better transport into the entrance of the perforations than water does.
6. The slurry should be pumped so that it dehydrates in the perforation tunnels.

References

1. Penberthy, W.L. Jr. and Cope, B.J.: "Design and Productivity of Gravel Packed Completions," *JPT* (Oct. 1980) 1679-86.
2. Torrest, R.S. and Savage, R.W.: "Particle Collection From Vertical Suspension Flows Through Small Side Ports—A Correlation," *Cdn. J. Chem. Eng.* (Dec. 1975) **53,** 699-703.
3. Gruesbeck, C. and Collins, R.E.: "Particle Transport Through Perforations," *SPEJ* (Dec. 1982) 857-65.
4. Penberthy, W.L. Jr.: "Gravel Placement Through Perforations and Perforation Cleaning for Gravel Packing," *JPT* (Feb. 1988) 229-36; *Trans.*, AIME, **285.**
5. Matherne, B.B. and Hall, B.E.: "A Field Evaluation of a Gravel-Diverted Acid Stimulation Prior to Gravel Packing," paper SPE 19741 presented at the 1989 SPE Annual Technical Conference and Exhibition, San Antonio, Oct. 8-11.

SI Metric Conversion Factors

bbl	× 1.589 873	E−01	=	m^3
cp	× 1.0*	E+00	=	mPa·s
ft	× 3.048*	E−01	=	m
gal	× 3.785 412	E−01	=	m^3
in.	× 2.54*	E+00	=	cm
lbm	× 4.535 924	E−01	=	kg

*Conversion factor is exact.

Chapter 8
Gravel-Pack Placement

Introduction

Gravel placement involves those operations required to transport gravel from the surface to the completion interval to form a downhole filter that will permit the flow of fluids into the well but will prevent the entry of formation sand. The gravel pack must be placed so that a uniform pack with a porosity of 38 to 39% or lower results. The gravel should also be placed so that the pack is not damaged to the point that well productivity is restricted. Low- or high-viscosity fluids can be used with several placement techniques to transport the gravel to the completion interval. The fluid and technique selected depend on well conditions, the probability of success, and the cost.

Choice of Fluids

Selecting a gravel-placement technique for completing a well usually involves choosing fluids to transport the gravel slurry to the completion interval. Water- and oil-based fluids, emulsions, and foams are commonly used.[1] Clean fluids are essential. Depending on well pressures, high-density, solids-free soluble salt solutions may be required to maintain well control. These fluids have densities as high as about 18 lbm/gal. The use of the high-density brines involves significantly higher completion-materials costs and requires handling procedures that necessitate additional rig time.

Table 8.1 lists some fluids used to gravel-pack wells. Most of these fluids can be further viscosified by adding polymers (see Fig. 7.10). The viscosities of the brine solutions also depend on the salt concentration and temperature. In many situations, brines are combined to achieve a given density at a reduced cost. The implication of the wide viscosity range is that higher-viscosity fluids have a greater ability to suspend gravel particles. Low-viscosity fluids can transport only about 1.0 lbm/gal (1.0 lbm of gravel in 1 gal of fluid—i.e., the total volume is greater than 1 gal) of gravel at normal pumping rates; the very-high-viscosity fluids, particularly crosslinked polymer solutions, can transport gravel at concentrations exceeding 15 lbm/gal. A thorough understanding of the properties and capabilities of these fluids can be critical to completion success.[2-9]

There are two techniques for placing gravel around a screen.[8-15] With conventional packing techniques, the gravel is placed at concentrations less than 2 lbm/gal by use of low-viscosity fluids. In slurry packing, high-viscosity fluids containing high gravel concentrations are used (4 to 15 lbm/gal). The delineation between the two is not always as simple as low vs. high viscosity because their uses overlap considerably. Different gravel-placement procedures are required when these two fluid systems are pumped. In some cases, gravel slurries are pumped by modified conventional practices to place the gravel. But because slurry-pack fluids have high viscosities, they can also suspend gravel and squeeze it through perforations so that the prepack and the gravel pack can be performed simultaneously. This option does not ordinarily apply to gravel placement with low-viscosity fluids. Here, perforation prepacking is commonly performed before the screen is run in the well. Regardless of the fluid chosen as the transport fluid, gravel placement ideally progresses from the bottom of the completion interval upward. Because of the well deviation and the fluid viscosity, this progression does not always occur. Thus, a thorough understanding of the gravel-placement mechanics of each fluid system is important in achieving the desired result—a sand-free, high-productivity completion.

Choice of the gravel-transport fluid is in many cases not the important issue. Conventional and slurry packing methods can produce completions with excellent productivity, provided that the proper method is selected for the particular application and that the gravel is packed properly in the perforations and around the screen. Regardless of the fluid system chosen, the materials, equipment, and placement techniques should be selected that will yield a sand-free completion with the desired productivity.

Conventional Packing

Conventional packing, sometimes called circulation gravel packing, normally involves the placement of gravel suspended in a low-viscosity transport fluid pumped at low gravel concentrations. Circulation packing is usually conducted after the perforations have been prepacked with gravel or in an open hole. Typically, the transport fluid is filtered brine with the gravel added at a concentration of 0.5 to 1.0 lbm/gal. The gravel is commonly mixed into the fluid through a gravel injector (shown later) while pumping at rates from about 0.5 to 3.0 bbl/min. Conventional packing is compatible with essentially all the placement techniques that are discussed later. **Fig. 8.1** is a schematic of gravel placement with conventional packing techniques and crossover equipment at well deviations between 0 and 45° from vertical that shows that, within this range of well deviation, the final result is the same. Gravel is transported into the annulus between the screen and casing (or open hole) where it is packed into position from the bottom of the completion interval upward. The transport fluid then returns to the annulus through the wash pipe inside the screen that is connected to the workstring. The

TABLE 8.1—PROPERTIES OF GRAVEL-PACK FLUIDS

Fluid	Approximate Density (lbm/gal)	Approximate Viscosity (cp at 60°)
Fresh water	8.33	1
NaCl brine	8.33 to 9.6	1 to 3
$CaCl_2$ brine	8.33 to 11.6	1 to 20
KCl brine	8.33 to 9.8	1 to 2
$CaBr_2$ brine	8.33 to 15.1	1 to 30
$NaBr_2$ brine	8.33 to 12.7	1 to 20
$ZnBr_2$ brine	8.33 to 19.0	1 to 20
Oil	7.0	1 to 100
Emulsions	8.33 to 11.0	5 to 100
Foams	0.3 to 4.0	1 to 50

wash pipe forces the fluid/gravel mixture to flow initially around the bottom of the screen. An increase in pump pressure is observed when the gravel level reaches the telltale screen. No further gravel placement is required above this point in the well. Telltale screens range from about 5 ft in length for short completion intervals to 30 ft when long intervals are gravel packed. Slot densities in telltale screens may be less than that of the gravel-pack screen. Some operators do not use upper telltale screens and extend the screen higher in the well.

The amount of blank tubing between the screen and the telltale (see Fig. 8.1) is called the gravel reserve. Its purpose is to replace gravel in the gravel pack that may be lost owing to settling over the life of the completion. When low-viscosity fluids are used, the gravel-pack reserve volume is usually equal to that opposite the screen but may vary depending on well conditions and the length of the completion interval; i.e., short intervals usually have a larger gravel reserve than long intervals.

Conventional gravel packing has both advantages and disadvantages over slurry packs.

Advantages.

- Positive indication of gravel level at gravel-pack completion.
- No mixing of complex fluids (usually).
- Low gravel concentrations.
- Minimum settling after placement.
- Better suited for long intervals and high-angle wells (see discussion of deviated wells).

Disadvantages.

- Long placement time may be required in long intervals.
- Possibility of screen erosion during placement.
- Requirement of a prepack if water is used as the carrier fluid.
- Possibility of gravel-size segregation.
- High fluid loss in certain situations.

Slurry Packing

Slurry packing[7,8] involves pumping gravel at high concentrations in a viscous transport fluid. The slurry is usually mixed in a blender or paddle tank before it is pumped. Although the fluid can be either water- or oil-based, hydroxyethylcellulose-viscosified brine is the usual choice. The typical polymer loading is 60 lbm/1,000 gal. In some situations, a carboxymethyl hydroxyethylcellulose fluid is crosslinked to create very high viscosities. The concentrations of gravel usually pumped with slurry-pack fluids is about 10 lbm/gal. In certain situations, however, concentrations have ranged from much less than 10 to more than 18 lbm/gal. When these high-density fluids are pumped, the fluid and gravel tend to move as a mass. Compared with the low-density conventional gravel-pack fluid, these slurry systems have significantly greater gravel-suspending capabilities. Pump rates normally range from ½ to 5 or 6 bbl/min when gravel is placed around the screen and depends on well conditions. **Fig. 8.2** also shows the gravel sequence with a viscous fluid at well deviations between 0 and 45° from vertical.

Slurry-pack gravel placement is also compatible with most gravel-pack equipment and techniques. When high-viscosity fluids are used to transport gravel into the completion interval, a reserve volume should also be specified. It usually is higher than required for low-

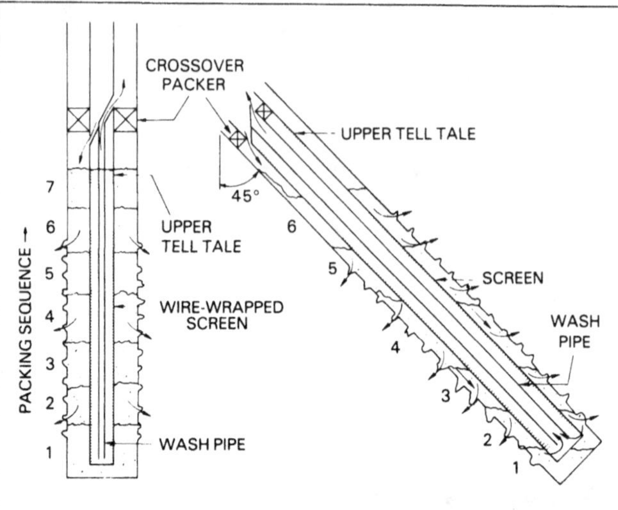

Fig. 8.1—Conventional packing techniques, crossover equipment.

viscosity fluids to allow for subsequent settling. Well conditions affect the gravel reserve volume when viscous fluids are used. Before the gravel is pumped, calculations should be made to determine the theoretical amount of gravel required to pack the annulus around the screen in addition to the gravel reserve. The placement should be about 100% under ideal conditions. Depending on the additional gravel placed through the perforations, possible bridging in the tubulars, or pumping conditions, the amount of gravel placed could differ from the theoretical volume. If it is considerably less, premature bridging probably has occurred and measures should be taken to settle the gravel around the screen and to pump additional gravel before the well is placed on production.

When high-viscosity fluids are used to circulate gravel around the screen, care should be exercised to pump at a low rate so that the viscous-drag forces do not exceed the gravitational forces on the gravel once the slurry has reached the completion interval. Should the viscous forces dominate, the gravel will be drawn into the screen rather than settling to the bottom of the well, as Fig. 8.2 illustrates. This causes the gravel to pack radially outward from the screen to the point that there may be premature indications of a completed gravel pack. This type of packing geometry and sequence is not desirable because, for the best packs, the gravel should dehydrate from the bottom of the completion interval upward. The potential for nonuniform packing is one of the disadvantages of slurry packs. Again, attention should be given to comparing the actual volume of gravel pumped to the theoretical volume. If the pumped gravel volume is much less than the theoretical volume when high-viscosity fluids are used, pumping should be stopped until the gravel settles before additional gravel is pumped to complete the gravel pack.

The use of an upper telltale is discouraged when high-viscosity fluids are used to circulate gravel to avoid bridging. The reason is that a gravel screenoff on the upper telltale is highly probable when higher-viscosity fluids are used. This screenoff could prevent gravel placement over the main completion interval. Recommendations are to extend the screen or slotted liner above the completion interval before blank tubing is added to achieve a gravel reserve when no upper telltale is used. Because of gravel settling, a portion of the gravel reserve volume will be filled provided that well deviations do not exceed about 60°.

A lower telltale screen is sometimes used with slurry packing; however, some operators do not use a lower or upper telltale with this fluid system but tube the wash pipe to the bottom of the screen section. Although the benefits of a lower telltale may be limited and it use complicates field operations, many operators use it. If a lower telltale is used, the wash pipe can initially be packed off above the lower telltale when crossover equipment is used (see Fig. 8.2). The lower telltale is claimed to ensure that gravel is initially

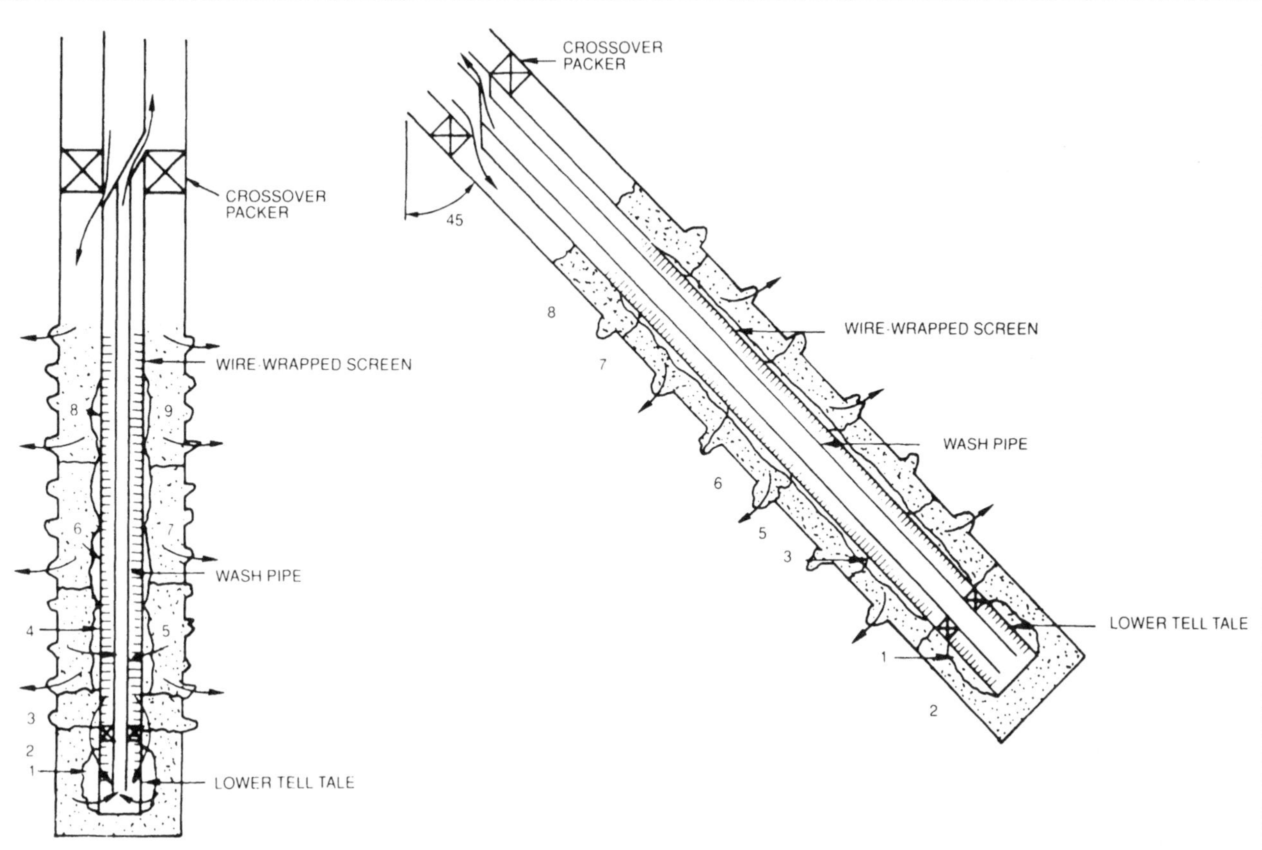

Fig. 8.2—Gravel-packing sequence with viscous fluids.

packed in the rathole. Hence, it is preferred that the screen be set as close as possible to the plugback total depth to avoid a large rathole interval. Because of the high fluid viscosity, the gravel settles slowly in slurries. Consequently, placing the gravel initially in the rathole region will avoid subsequent disruption of the gravel pack after placement by settling. After the rathole is filled, gravel can be squeezed into the perforations. When the wash pipe is removed from the packoff, circulation can be continued through the wash pipe above the packoff to complete the gravel pack around the screen. Hence, the entire prepack and gravel pack can be completed in a single operation. Also, because high gravel concentrations are used, the gravel pack can be performed quickly compared with low-viscosity conventional packs.

Viscous fluids provide excellent gravel transport into the entrance of the perforations when the gravel descends down the well compared with water because the drag forces on the gravel are high. The perforation-packing efficiency in the perforation tunnels is high once the gravel enters the perforation, regardless of the fluid used. However, there must be fluid loss for high perforation-packing efficiency, as mentioned previously.

When slurries are pumped, it is important to know the approximate viscosity of the fluid at downhole conditions[3,4] to enhance placement efficiency. Viscous slurries are sometimes difficult to dehydrate if premature bridging occurs in the completion interval. The result is that voids may form in the gravel pack—e.g., adjacent to blank sections, couplings, and other locations in the gravel pack—particularly in long or long, deviated intervals,[5] and will remain unless the bridges break down with time and allow the gravel to settle. Another disadvantage of slurry packs is the potential for nonuniform packs that initially contain voids that may not be removed by settling. This occurrence is uncommon when water is used as the placement fluid.

Although desirable, complete dehydration of the gravel slurry may not be essential when low-deviation wells are slurry packed. Viscosity break and subsequent settling can also create a gravel pack,[6] as **Fig. 8.3** shows. However, sufficient gravel volume must be available in the gravel reserve to obtain a settled pack without

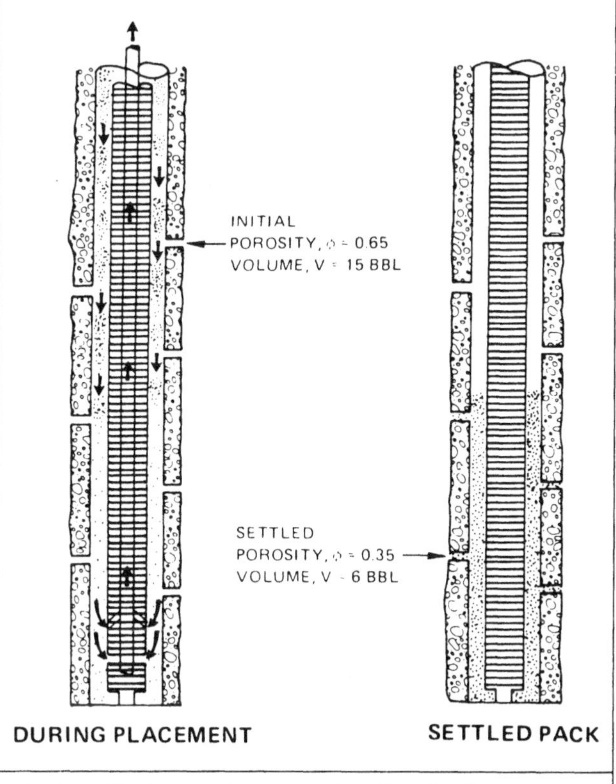

Fig. 8.3—Displacement process when first significant amount of slurry reaches bottom.

exposing some of the completion interval to the screen. The settled volume of gravel whenslurries are pumped is related to gravel concentration. The settled bulk gravel volume of 10 lbm of gravel in 1 gal of neat slurry is about 0.45 gal; 15 lbm of gravel in 1 gal of neat slurry settles to a bulk volume of about 0.65 gal. If settling

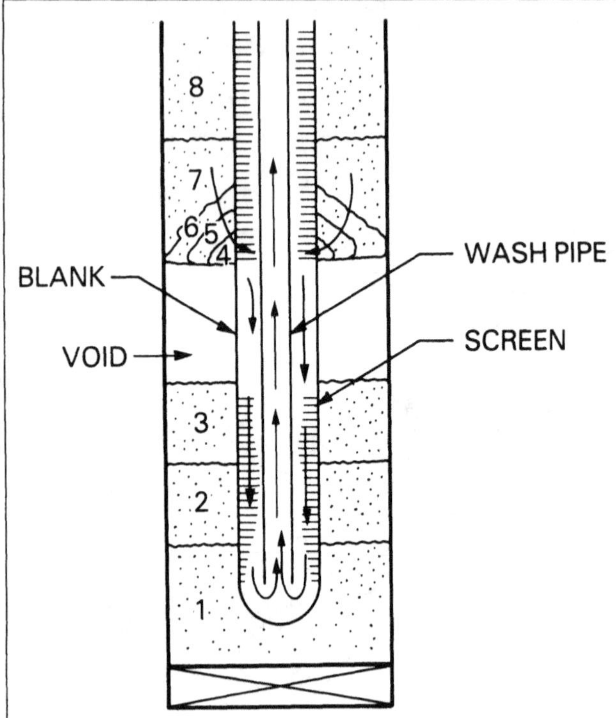

Fig. 8.4—Void formation opposite blank sections caused by viscous fluids.

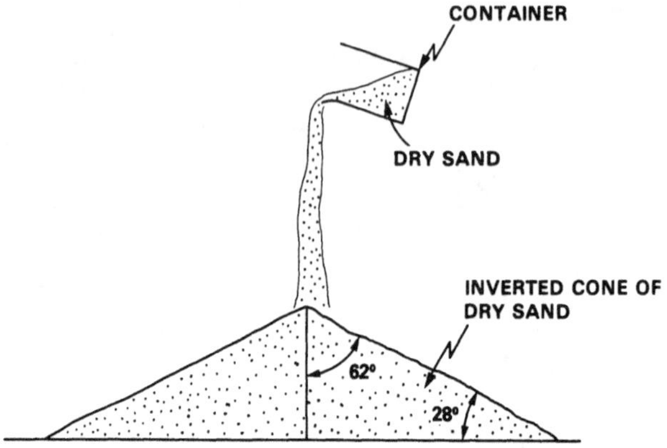

is relied upon to complete the gravel pack, large gravel reserve volumes may be required compared with conventional packing. In short intervals, a gravel reserve several times the completion interval is not uncommon when slurries are pumped. After settling, about one-half this volume will contain gravel. To accelerate settling after placement, polymer-viscosified fluids should always be accompanied by a breaker to reduce the fluid viscosity by 90% or more after a few hours at the formation temperature. While not preferred over dehydration during placement, settling is an alternative, but it probably results in higher pack porosities. Before the well is produced, however, complete viscosity break is essential to provide a low-porosity gravel pack.

Attention should be given to the screen assembly so that no blank sections are included. Blank sections can cause voids in the pack opposite these intervals when viscous fluids are used. The void illustrated in **Fig. 8.4** is the consequence of the viscous forces dominating the gravitational forces. This results in voids opposite the blank sections as a consequence of bridges that form at the top of the blank section. Hence, blank sections should be avoided when gravel packs are conducted with viscous fluids.

CONTAINER
DRY SAND
INVERTED CONE OF DRY SAND
62°
28°

Fig. 8.5—Angle-of-repose schematic.

A problem that must be contended with is slurry roping in the tubing caused by its high density compared with those of the fluids in the tubing. This difference in density causes the gravel to settle in mass ahead of the transport fluid. This condition exists only at low pump rates or when a low-viscosity fluid is ahead of the slurry. When it occurs, it could lead to premature indications of gravel arrival at key locations in the well. To minimize roping, a lead pad of gravel-pack fluid can be pumped ahead of the slurry. Also, pumping the slurry at velocities of about 500 ft/min while the slurry is in the tubing will minimize roping. When the slurry arrives at the completion interval, pump rates usually should be reduced when gravel is placed around the screen to reduce premature bridging of the gravel high in the completion interval.

Use of slurry packs has both advantages and disadvantages.

Advantages.

- Shorter pumping time.
- Low fluid-volume requirements (and tankage).
- Reduced pumping time because of high sand concentrations.
- Simultaneous prepacking and gravel packing.
- Minimum returns through the screen, which may reduce erosion.

Disadvantages.

- No positive indication of the gravel level around the screen when placement is ended.
- Nonuniform gravel packs.
- Several joints of blank pipe to allow for uncertainties in placement; the gravel reserve is sometimes well over 100%.
- Requirement of complex fluids.
- Potential bridging in blank sections of the screen.
- Problems with gravel dehydration owing to slow gravel settling.
- Not well suited for long, highly deviated completion intervals with low-viscosity transport fluids (see Gravel Placement in Deviated Wells).

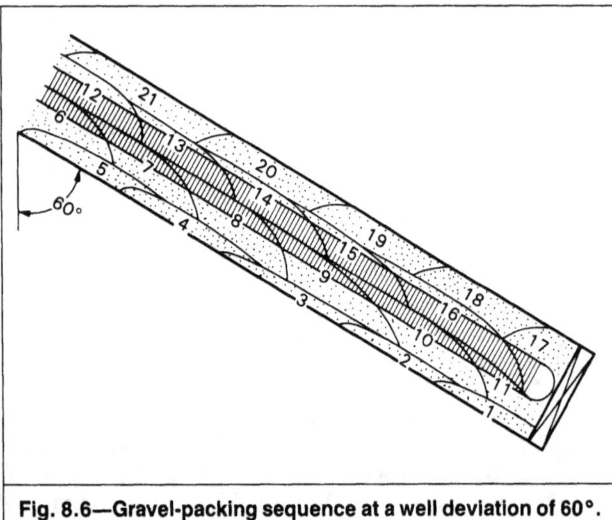

Fig. 8.6—Gravel-packing sequence at a well deviation of 60°.

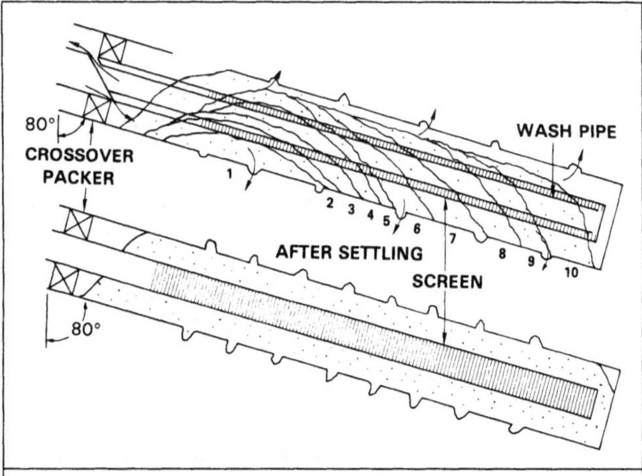

Fig. 8.7—Gravel-packling sequence in an 80° deviated well with large-diameter washpipe (ID/OD ratio = 0.8).

Probably the most attractive feature of slurry packing is that it can be performed very quickly owing to high gravel concentrations. Another principal advantage is that prepacking and gravel packing the completion interval can be performed simultaneously, saving thousands of dollars in rig costs. Slurry packing does have limitations, however. Dehydration problems occur with slurries. Also, their use in highly deviated wells may be ineffective if the pack is not effectively dehydrated during placement and pack settling creates voids.

Gravel Placement in Deviated Wells

Model studies[10-14] have indicated that effective gravel placement in deviated wells can be achieved with a rather wide range of fluid viscosities but requires special placement-design considerations. But regardless of the fluid viscosity and placement techniques used, the slurry must be dehydrated so that a low-porosity gravel pack exists upon cessation of pumping operations.

Changes in gravel transport begin when the well deviation exceeds about 45° but become critical at deviations above 55 to 60°. The reason is that gravel's angle of repose is about 25 to 28° and represents the angle between the horizontal and a stable inverted cone of gravel that is in equilibrium with a flat surface. The angle of repose is the complement of the well deviation. **Fig. 8.5** illustrates this concept. The implication of this information is that effective gravel packs should be possible up to well deviations of 62 to 65° without special designs; however, this analysis does not take into account boundary effects from the casing and screen. Hence, gravel packs at well deviations greater than 60° will settle to form a void over the top of the pack unless dehydration is complete and the gravel is at its lowest possible porosity when pumping is completed. **Fig. 8.6** shows the gravel-packing sequence at a well deviation of 60° and illustrates the complexity of the placement because the gravel deposition is taking place at about the angle of repose.

Conventional Packing. Properly executed gravel placement in deviated wells (greater than 60°) with low-viscosity fluids initially results in gravel-dune formation[11,14] on the low side of the well high in the completion interval. As additional gravel enters the screen annulus, the dune height increases to an equilibrium height controlled by the pump rate. Any subsequent gravel transport occurs over the top of the dune with deposition on the back side. Thus, the gravel dune lengthens until it reaches the lower portion of the screen or rathole. At this point, the channel over the top of the settled gravel is packed. **Fig. 8.7** illustrates this packing sequence.

For gravel to be packed effectively at the high angles (60° or more), the gravel-pack equipment and procedures must be analyzed critically. One of the main problems during low-viscosity placement is that the gravel dune tends to divert fluid into the well screen. If too much fluid is diverted, there is insufficient flow in the outer-screen annulus and the gravel bank ceases to proceed downward.

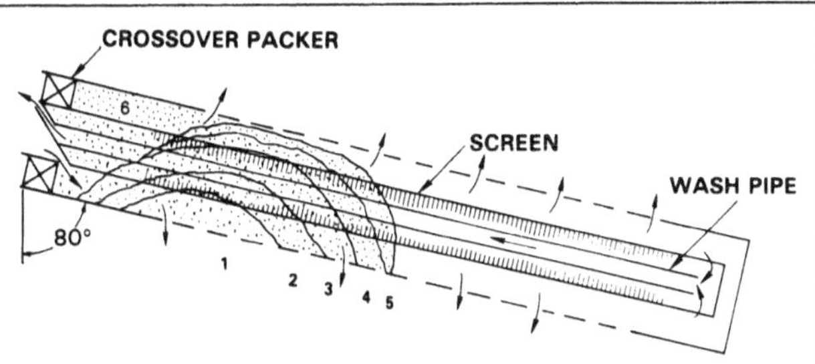

Fig. 8.8—Stalled gravel pack caused by fluid diversion into the screen/washpipe annulus (ID/OD ratio = 0.5).

A sandout high in the completion interval and insufficient gravel placed around the screen results. **Fig. 8.8** also shows this packing situation.

Two techniques have been used to avoid this occurrence: using cups on the washpipe[16] or increasing the wash-pipe diameter.[11] Both techniques divert flow to the screen/casing annulus, resulting in superior gravel placement. For operational reasons, increasing the wash-pipe diameter is the preferred technique because it increases the resistance to flow between the screen ID and the wash-pipe OD. This diverts fluid to the outer-screen annulus to transport the gravel down the well. Although the wash-pipe-OD/screen-ID ratio should be 0.8 or higher for the best results, because of clearance problems in the well, wash pipe with a ratio of about 0.8 seems to be the largest-diameter pipe that most operating personnel are willing to run. While these high diameter ratios are necessary for high-angle wells, they also assist in packing vertical wells. **Fig. 8.9** compares the placement for two different wash-pipe sizes and shows that the use of a large-diameter wash pipe maintains high placement efficiency. High-angle wells have been successfully gravel packed with the large-diameter wash-pipe approach when only water was used as the transport fluid.

Increasing only the wash-pipe diameter does not ensure effective placement. Maintaining a high pump rate is also quite important for efficient placement in deviated wells.

Unlike gravel placement in vertical wells, gravel packing in highly deviated wells depends on the pump rate when low-viscosity fluids are used. Pump rates should be as high as possible as long as excessive bottomhole pressures are not generated. For effective placement, the superficial velocity (ratio of the pump rate to the annulus area around the screen) should be at least 1 ft/sec to avoid the formation of a premature bridge high in the completion interval.

Gravel packing highly deviated wells with conventional gravel-pack equipment does not pack the rathole very well. When circulation-type equipment is used, the gravel-pack assembly should be placed as low in the well as possible to pack the rathole. Also, the use of an upper telltale screen has no benefit and can even prevent

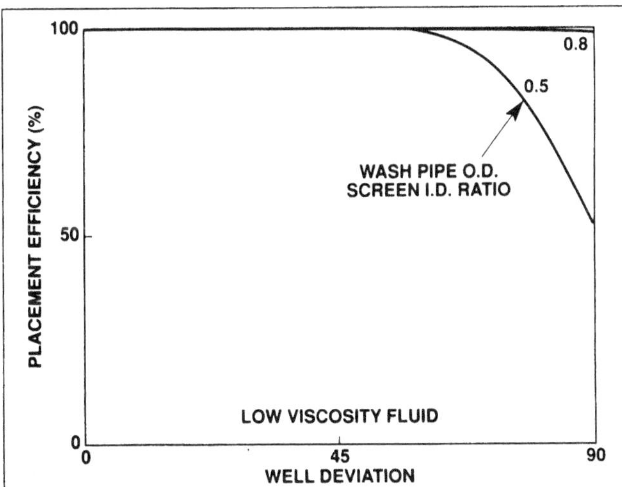

Fig. 8.9—Gravel-pack efficiency as a function of well deviation and wash-pipe size.

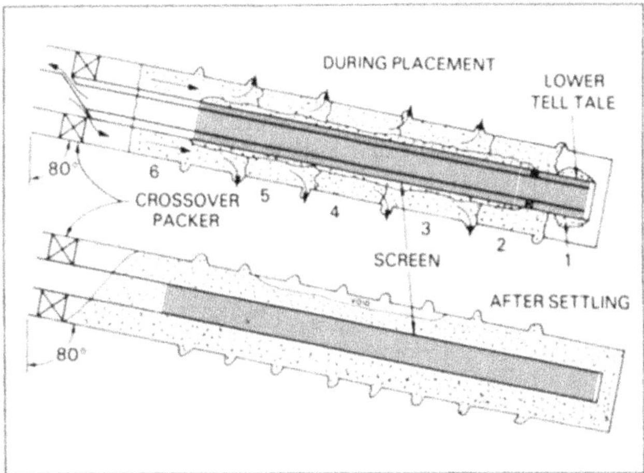

Fig. 8.10—Gravel-packing sequence with viscous fluids and void formation after settling.

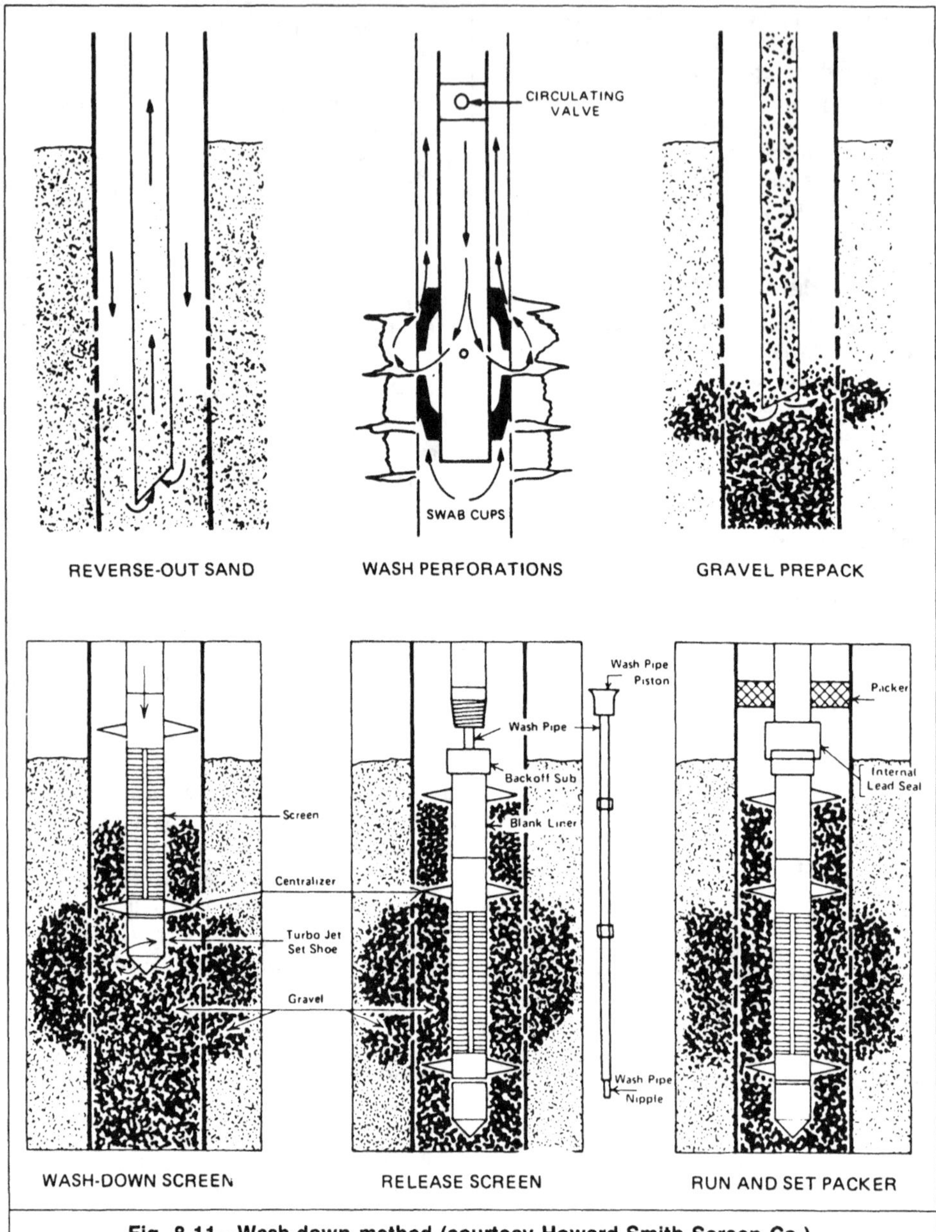

Fig. 8.11—Wash-down method (courtesy Howard Smith Screen Co.).

effective gravel placement if well angles exceed 45°. They should be omitted in these situations.

An advantage of using low-viscosity fluids for highly deviated wells is that, after the completion interval has been effectively packed, little or no settling occurs. Also, no attempt should be made to settle the gravel in high-angle completions. Voids can form across the top of the pack and may lead to gravel-pack failure. Some operators[17] advocate perforating the low side of the well only in these situations.

Slurry Packing. In highly deviated wells, gravel is transported with viscous fluids completely differently than with low-viscosity carrier fluids. In this case, higher gravel concentrations (about 10 lbm/gal or more) are usually used in fluids that contain polymer loadings of 60 to 80 lbm/1,000 gal as are used in pumping slurries in vertical wells. These fluids provide excellent proppant transport. Gravel tends to be transported in mass along the low side of the well because of the high gravel concentration and the high fluid viscosity.

When viscous fluids are pumped with crossover gravel-pack equipment, a lower telltale assembly is usually used. In fact, an upper telltale is not recommended and may cause a premature bridge at that location. The initial flow of fluid is through the lower telltale, causing gravel to be packed around it. Once packing is completed across the telltale, the gravel-pack equipment should be manipulated to squeeze fluid through the perforations to prepack them before diverting returns through the wash pipe in the upper circulating position to gravel pack around the screen. Unlike the low-viscosity fluids, high-viscosity fluids move gravel from the lower telltale screen upward.

Gravel dehydration in wells deviated more than 55 to 60° becomes more difficult with slurries than with low-viscosity fluids. Studies[18] have shown that dehydration usually is good at the lower and upper portions of the gravel-packed interval but that a void exists along the top of the middle of the pack (see **Fig. 8.10**). This void is created because the viscous forces dominate the gravitational forces. Consequently, the gravel bridges prematurely at the top of the interval before the middle of the gravel pack has time to dehydrate. Thus, dehydration is good at the top and bottom of the gravel pack, but voids exist at the top of the pack in the intermediate locations. Hence, gravel packs performed with slurry-pack techniques require special attention to achieve acceptable placement and dehydration in highly deviated wells.

For the best pack efficiency with slurries, the gravel pack should also be conducted with a wash-pipe-OD/screen-ID ratio of 0.8. This practice has benefits even in vertical wells. Pumping at a high rate is not as critical for effective placement with viscous fluids as with low-viscosity fluids. Dehydration can be significantly improved when slurries are pumped by reducing the gravel concentration to about 3 lbm/gal or less.

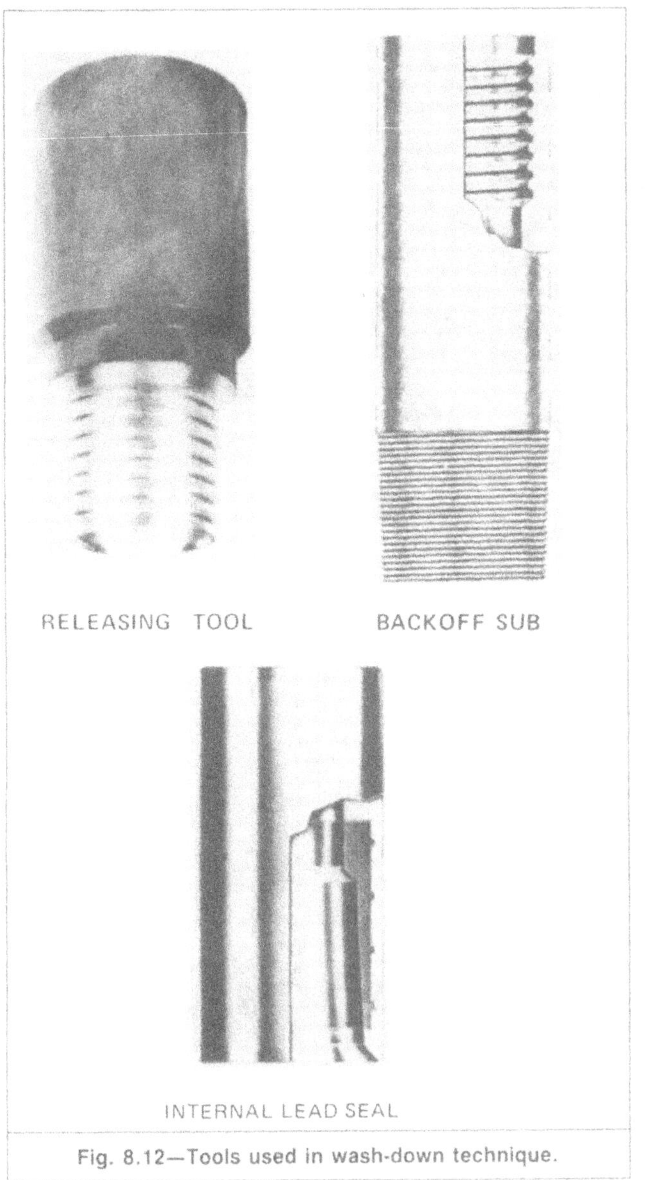

Fig. 8.12—Tools used in wash-down technique.

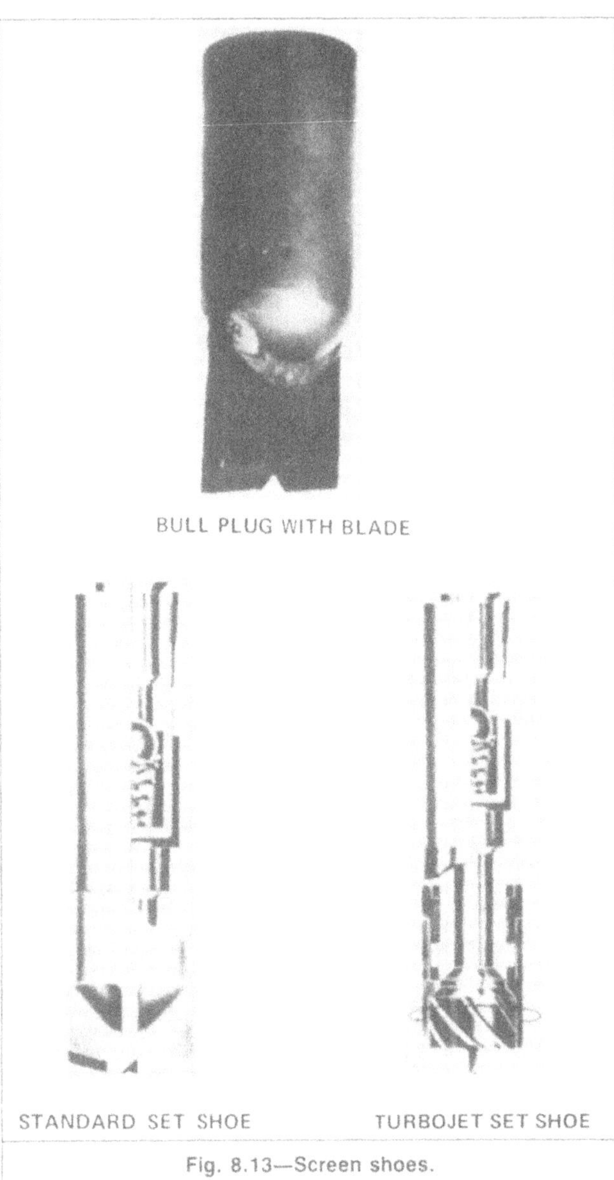

Fig. 8.13—Screen shoes.

Techniques and Equipment for Gravel Placement

Several techniques and types of equipment are available for gravel packing wells. Each technique has advantages and limitations. The most successful gravel-pack techniques are the wash-down, reverse-circulation, and crossover-placement methods. Most of the other placement methods are modifications of these techniques. Service companies may also use slightly different equipment and procedures to achieve the desired gravel placement. All these placement techniques can be used for either open or cased holes.

Wash-Down Method. The wash-down method (**Fig. 8.11**) is a simple completion technique but is performed quite differently from the placement methods described previously. As in all cased-hole gravel-pack operations, it is necessary to clean the wellbore and the perforations and then pack the perforations with gravel. The completion sequence is as follows.

1. Once the perforation prepack is completed, gravel is left inside the casing at some predetermined height.
2. The screen is run to the top of the gravel, and circulation is initiated through the bottom of the screen. **Fig. 8.12** shows several devices that allow sufficient circulation to fluidize the gravel.
3. The screen is washed into place, circulation is stopped, and the gravel is allowed to settle around the top of the screen or slotted-liner section.
4. The working string is released from the screen with a backoff sub and is raised a few feet.
5. Reverse circulation is initiated up the wash pipe to ensure that the inside of the screen is clean.
6. When necessary, a second trip is made to put a packer device on top of the screen assembly.

The major disadvantage of this method is that the length of screen that can be washed down successfully is limited. Discontinuing circulation to make a connection can cause gravel to settle around the screen and may prevent further screen movement. As a result, the maximum screen setting is usually no more than one joint of tubing if the rig can handle single strings only. If a larger rig is available, the length of the screen setting is limited to the length of a stand of pipe.

Either low- or high-viscosity fluids can be used with this technique. The only requirement is that the gravel be sufficiently fluidized so that the screen can be washed into position.

Tools like those in **Fig. 8.13** (and Fig. 8.12) are needed for this type of installation. The releasing tool and back-off sub have left-hand Acme threads, which allow right-hand rotation for releasing from the top of the screen. The back-off sub is screwed on top of the screen assembly, and the releasing tool is part of the workstring. Fig. 8.12 shows an internal lead seal device that is run below a conventional packer and affixed to the top of the back-off sub to provide a seal between the two. The bull plug with a welded blade on the end in Fig. 8.13 provides some capability of washing down or rotating the pipe when necessary. Pipe rotation usually is not recommended because the screens are not designed to withstand high torque forces. The turbojet and standard check-valve shoes often are used in wash-down operations. Theoretically, the turbojet shoe is an improvement over the standard jet shoe for wash-down techniques because the lower blade assembly creates greater tur-

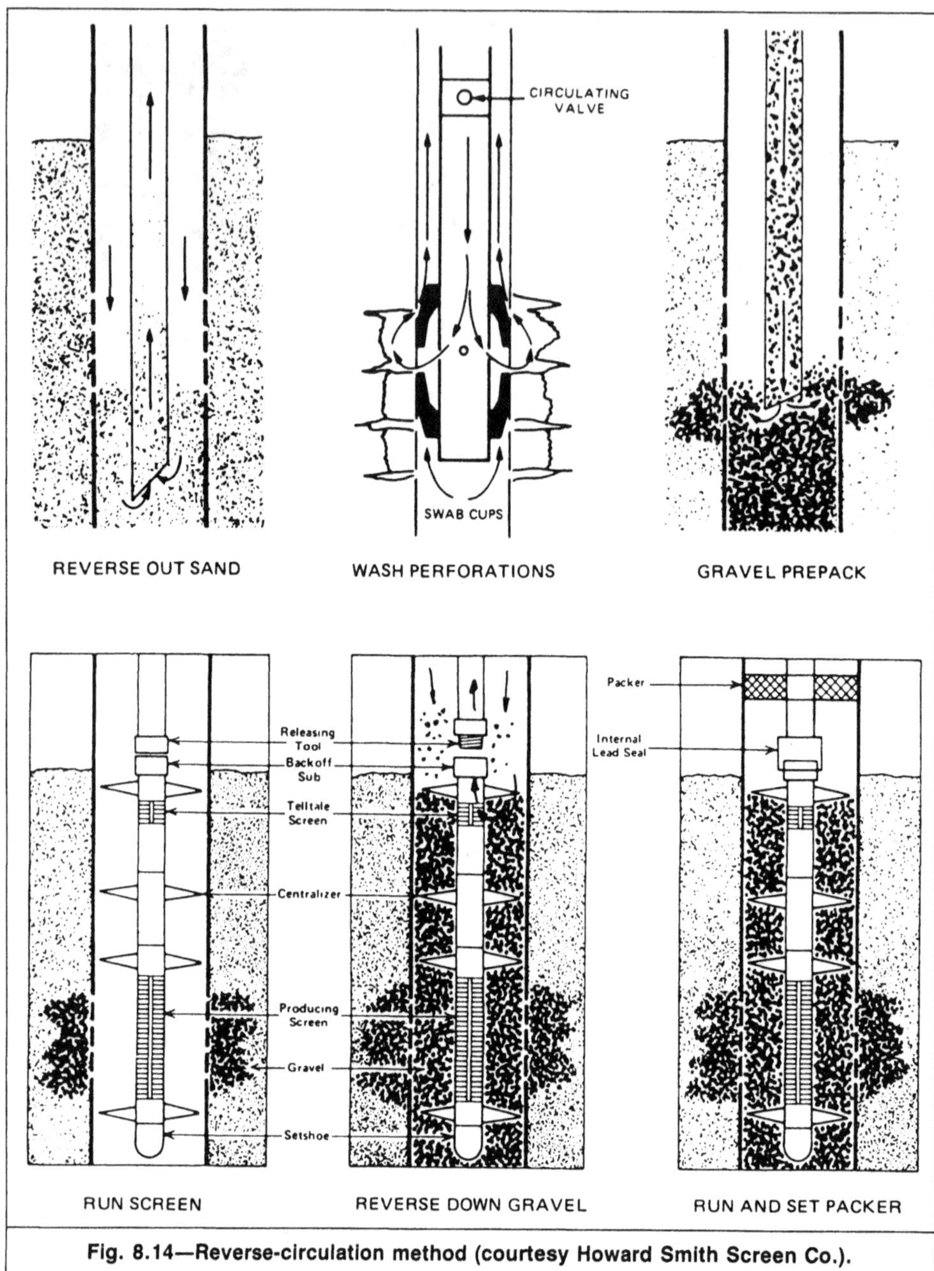

Fig. 8.14—Reverse-circulation method (courtesy Howard Smith Screen Co.).

bulence as the circulating fluid exits from the end of the wash-pipe assembly.

Reverse-Circulation Method. This method of gravel placement (see **Fig. 8.14**) requires wellbore preparation similar to that needed for the wash-down method, except all gravel is washed from the wellbore after the perforation prepack. The screen is then run and set in position, and the carrier fluid and gravel are reverse circulated to deposit the gravel around the outside of the screen. This method usually is used in conjunction with low-viscosity fluids, but high-viscosity fluids are acceptable when pumped at slower rates.

The releasing tool, back-off sub, and lead seal (Fig. 8.13) used in this technique are essentially the same as those used in the wash-down method.

The primary disadvantage of this technique is that the tubing and casing string may be contaminated with debris that could interfere with the gravel-pack permeability. Hence, cleanliness is a major concern.

Crossover Method. In the crossover gravel-pack method (**Fig. 8.15**), a workstring pumps gravel slurry around the screen. With this equipment, the slurry is crossed over between the screen and the casing at the top of the screen section. The gravel is then placed in the screen annulus while fluid returns to the surface through the well's annulus.

This technique is preferred for most situations. Crossover gravel placement has several advantages over wash-down and reverse-circulation techniques.

- The gravel slurry will not scour mud, rust, pipe dope, or scale from the casing. Thus, the risk of depositing debris in the perforations or on the screen is reduced.
- The workstring volume is smaller than that of the workstring/casing annulus. At equal pumping rates, the higher fluid velocity inside the workstring decreases placement time and the possibility of bridging or gravel segregation.
- Control of both fluid and gravel location within the workstring is more precise.
- Additional trips to run the lead seal as described for the previous techniques are circumvented because the packer provides a positive seal for the screen (only when crossover packers are used).

A disadvantage of the crossover method is that it is more expensive and may result in an expensive workover if the packer cannot be easily removed.

Crossover-type gravel placement can be accomplished with a liner hanger (**Fig. 8.16**), a cup-type crossover tool (**Figs. 8.17 and 8.18**) or a crossover gravel-pack packer (**Fig. 8.19**). With the cup-type equipment, the crossover tool is part of the workstring and consistes of a dual swab-cup assembly and crossover device. It is re-

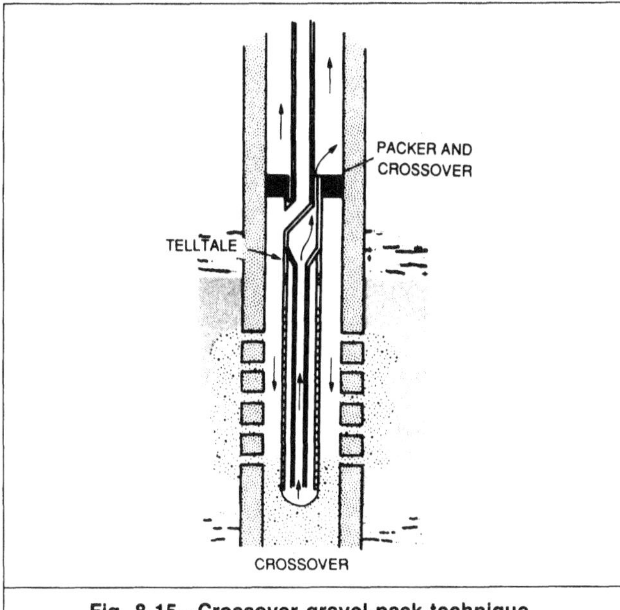

Fig. 8.15—Crossover gravel-pack technique.

moved from the well after gravel packing is completed. A second trip is required to set a packer or seal on top of the screen assembly.

Several service companies offer a crossover gravel-pack tool for use with one or more of their standard completion packers (Fig. 8.19). They generally use a ported extension with a crossover assembly attached to the production packer and extending through the ported extension. Additional equipment (such as wash pipe inside the screen with a backpressure valve) may also be added to the assembly.

The packer/crossover assembly has five major functions.

1. A means for setting and testing the packer.
2. A flow path for the sand slurry to be pumped down the workstring and out through the ports in the ported extension below the packer.
3. The capability either to circulate fluids through the screen and out the workstring/casing annulus (circulating position) or to bullhead the fluids into the formation (squeeze position).
4. A means to release the crossover tool assembly from the packer.
5. The capability to reverse circulate and clean out the workstring after releasing from the packer.

Before selecting a supplier of such tools and services, the operator should consider the following factors that affect the success of gravel-pack placement.

- Wellbore fluids should not be allowed to flow through the screen during tripping in the hole. Problems with screen plugging can easily occur in open holes, especially in old wells.
- The ports on the crossover-tool outlet and ported extension should be large and should overlap. These features are especially important when large gravel or high-sand-concentration slurries are pumped.
- Some means must be provided to close or pack off across the ported extension after completion of the pack.
- A positive indication of the tool position, such as squeeze or circulation, should be available. For example, a sudden increase in pickup weight may be a clear indicator of a correct position. (A collet is commonly used for this purpose.)
- There should be flexibility for use with either an upper-telltale or packed-off lower-telltale screen assembly.
- A squeeze position should be maintained without excessive tubing weight.

The various operations possible with crossover/packer assemblies are shown in Fig. 8.19. Neither an upper nor a lower telltale screen is illustrated. Fig. 8.19 does illustrate the use of a sump packer in combination with a gravel pack. The sump packer serves as a base for the screen and makes the rathole available for the settled solids in the well's flow stream. An alternative to the use of a sump packer is a bridge or cement plug as a base.

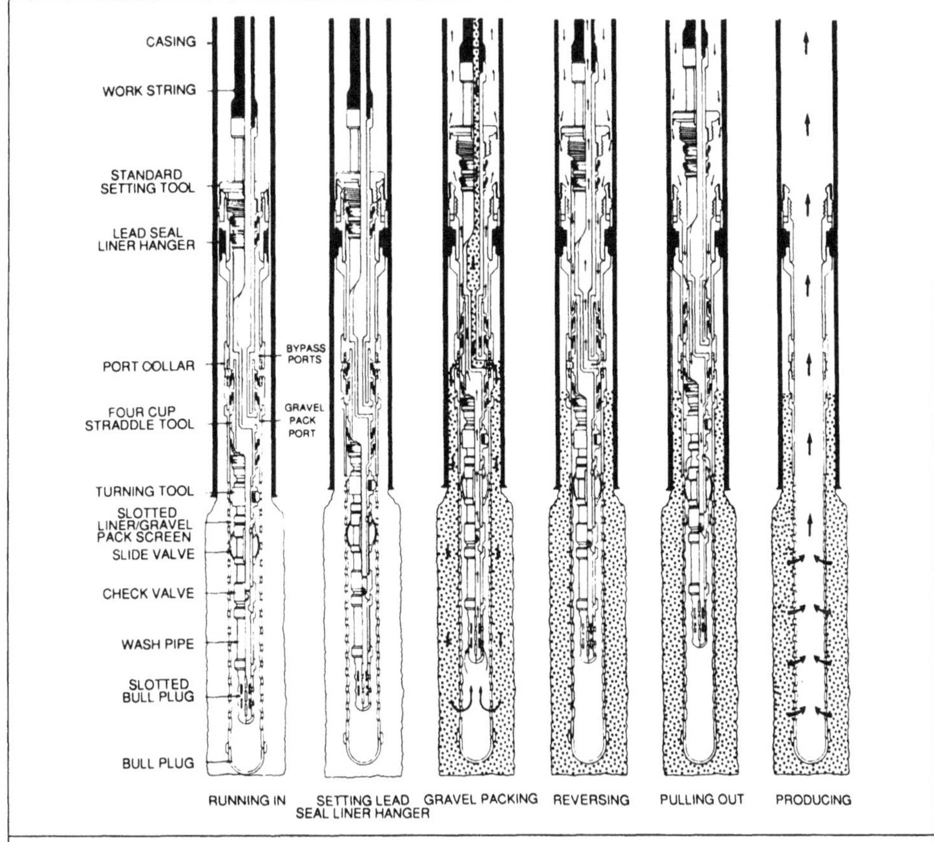

Fig. 8.16—Schematic of a linear hanger/crossover gravel pack (openhole) (courtesy Baker Sand Control).

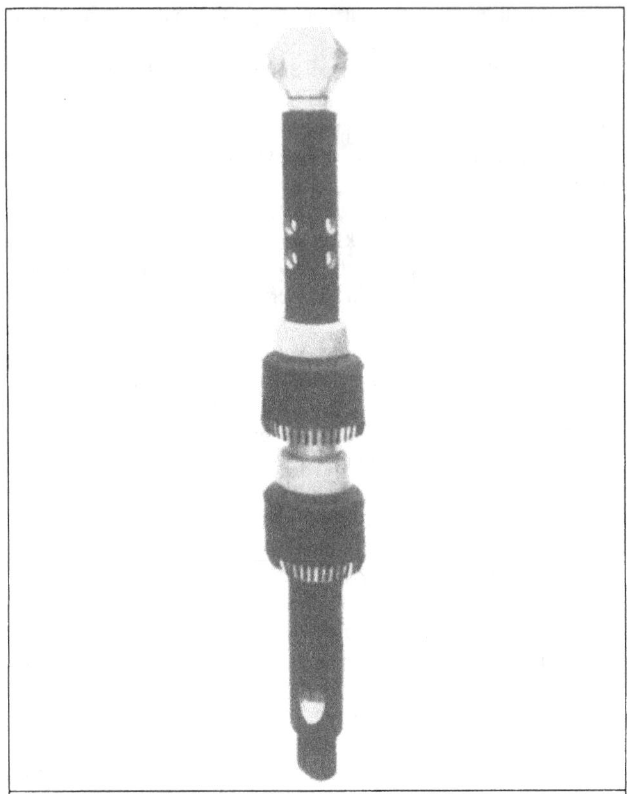

Fig. 8.17—Cup-type crossover gravel-packing tool.

Rathole annular volume below the screen should be the minimum possible, as mentioned previously. This is particularly important for good gravel placement in deviated wells and when viscous fluids are used. While the bottom of the screen assembly should be as close to the bridge or cement plug as possible, it should be positioned so that the screen is not in compression. The normal procedure for locating the screen in the well is to set the packer so that the screen bottom is less than 0.5 ft above the plug. If a sump packer is used, the screen is usually located in a collet that requires several thousand pounds of upward pull before it can be removed. After latching into the collet, the screen assembly is sealed at the top by setting the gravel-pack packer. The stress state of the screen at this point should be essentially neutral.

Centralization. In the installation of any gravel pack, regardless of the completion technique, the screen should be centralized so that gravel can be uniformly packed around it. Spring, weld-on, aluminum, and plastic centralizers are available (**Fig. 8.20**). The weld-on model, sometimes called a guide, is the most commonly used centralizer in cased-hole gravel packs. Some companies prefer not to weld to pipe and therefore choose clamp-on centralizers available in aluminum and plastic. Solid devices offer the best support for centralizing the screen inside the casing. The spring centralizers are usually run for centralization in openhole gravel packs. Centralizer spacing usually is between the screen sections at the coupling. In vertical wells, one centralizer every 30 ft is sufficient. In deviated wells, centralizer spacing becomes more critical. In wells where deviations exceed 60°, a 20-ft or closer centralizer spacing is advisable.

Other Gravel-Pack Equipment

In addition to filters and pumps, some means of mixing the gravel with the carrier fluid is necessary before it can be transported into the well. Gravel injectors, blenders, and paddle tanks are commonly used.

Gravel injectors (**Fig. 8.21**) can inject the gravel downstream of the pumps. An advantage of their use is that the sand is not flowed through the pumps, which is claimed to reduce crushing. Gravel injectors are usually used with very-low-viscosity transport fluids (water). Two limitations of their use are that regulating the gravel concentration over sustained periods of time is difficult and that they can be used only for low gravel concentrations. The gravel concentrations that can be sustained are about 0.5 to 1.5 lbm/gal.

Auger-type sand injectors, sometimes called gravel infusers (**Fig. 8.22**), provide a means of conveying gravel into the flow stream over a wide range of gravel concentrations, regardless of the viscosity of the fluids pumped. The infusers control the gravel concentration with an auger that conveys the gravel into the suction of the pumping unit at a uniform rate controlled by their rotational speed. The primary advantage of the infuser is that the gravel concentration can be conveyed more uniformly and precisely than conventional pot-type sand injectors, yet they have the flexibility to operate over a wide range of gravel concentrations. They also can handle bulk quantities of gravel, an impossible job for pot-type injectors because of their limited capacities.

Paddle tanks (**Fig. 8.23**) are commonly used for mixing slurries. These tanks have either hydraulic or mechanical (or both) mixing capabilities. The mixing can be on either a continuous or a batch basis. Paddle tanks are usually used for mixing slurry-pack fluids. After the fluid is prepared, the gravel is added. Gravel is suspended in the fluid by constant mechanical and/or hydraulic agitation, complemented by the high fluid viscosity. Depending on the fluid viscosity, gravel concentrations in excess of 15 lbm/gal can be mixed in these tanks. A disadvantage of premixing the gravel slurry is that the gravel must be conveyed through pumps before it is transported to the well. This increases the possibility of some disintegration. However, disintegration is not usually a problem.

Blenders are modifications of paddle tanks and provide a means of mixing gravel in a fluid stream and pumping it from the tank in a continuous operation despite the small blender volume (5 to 10 bbl). With this capability, premixing of gravel and fluid is not required, but the fluid may require mixing before it is inducted into the blender.

Completion Execution

Gravel-pack execution requires attention to details by personnel knowledgeable in all phases of the operation. This is a key requirement for a properly executed gravel pack. Remember that a gravel pack is nothing more than the installation of a downhole filter. Any operation that prevents or limits the gravel pack's ability to filter fluids effectively or restricts well productivity hampers an effective completion.

Before any gravel-pack equipment is run into the well, the completion engineer should be satisfied that the casing and workstring are clean; that clean, nondamaging workover fluids have been used; and that proper perforating or underreaming practices have been applied. (These items are discussed in Chap. 6.)

Considerable care should be taken during the handling of all gravel-pack equipment. The screens should not be uncrated until they are on the rig floor. They should be thoroughly inspected to ensure that they are undamaged and that all specifications have been met. When the screen is being "made up," the operator should be careful to avoid damage, particularly allowing the screen to contact the "spider," blowout preventers, or other objects on or near the rig floor. The screen should contain centralizers to minimize damage when it is run in the hole and to center it in the completion interval. The entire gravel-pack assembly should meet specific well conditions and should be spaced properly. The screen should overlap the completion interval by a minimum of 5 to 10 ft.

The screen and gravel-pack equipment should be run in the well slowly, with the gravel-pack packer set to prevent flow through the screen. This is usually the squeeze position on a crossover-type gravel-pack packer. Excess pipe dope is a source of potential formation damage when combined with particulates. Connections on the workstring do not require pipe dope for the trip to run the screen. If pipe dope is considered absolutely necessary, only the pins should be doped, and those sparingly. Similarly, no equipment run with the gravel-pack assembly should be painted because the paint could peel and cause contamination.

Attention to rate and pressure schedules is important during gravel packing. The gravel volumes pumped into the well (and reversed

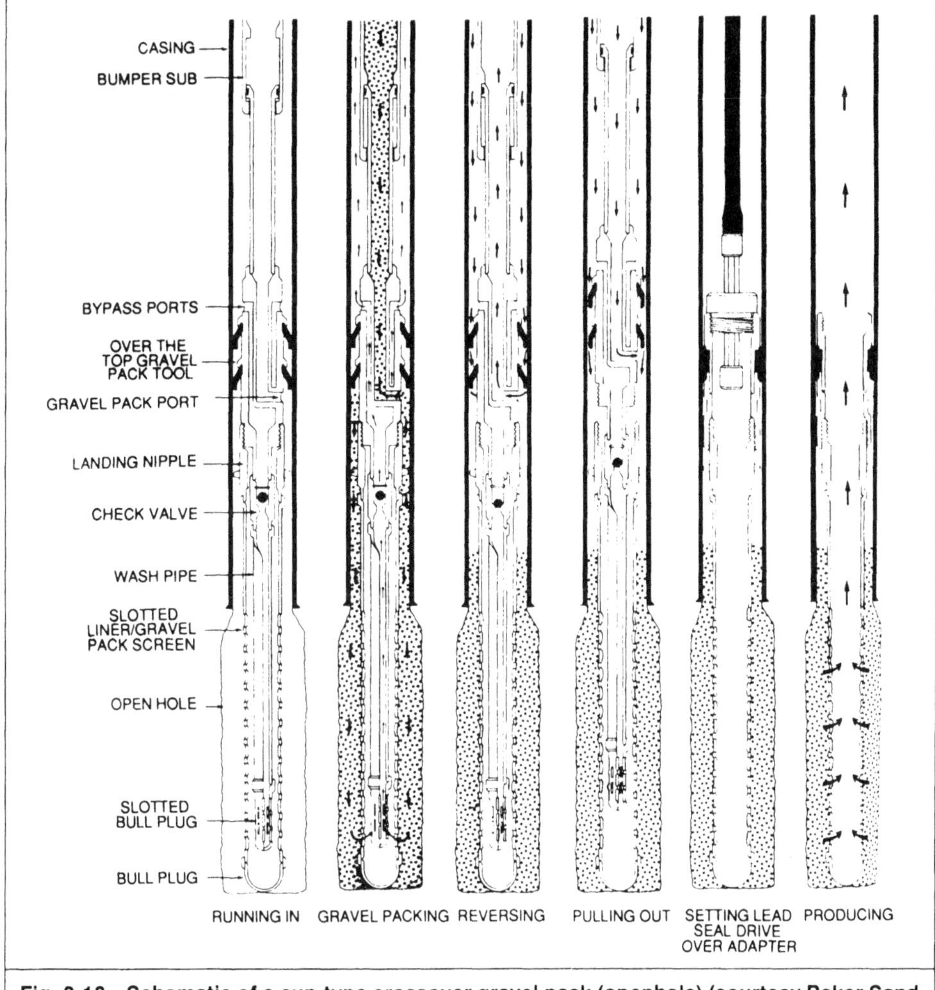

Fig. 8.18—Schematic of a cup-type crossover gravel pack (openhole) (courtesy Baker Sand Control).

out) should be compared with theoretical volumes. Actual pumped gravel volumes that exceed 100% of the theoretical volume are not uncommon. Gravel volumes significantly lower than the theoretical volume, however, probably indicate ineffective gravel placement, usually voids in the gravel pack or settling problems. In these situations, measures should be taken to correct the discrepancy. Sometimes this will require only the cessation of pumping to allow the gravel to settle in the completion interval. In other cases, more sophisticated measures may be required to settle the gravel. Gravel-pack washing should not be used to settle gravel or to eliminate voids. This practice is believed to promote gravel/formation-sand mixing.

Gravel-Pack Evaluation

Several logging techniques have been used to measure the degree of packing in the annular space around the screen.[19,20] These logging tools are capable of determining the porosity or density of the gravel pack. When they are run after a gravel pack, voids that could lead to completion failure can be identified. If voids are detected, the pack should be settled before the well is produced. In some cases, time will settle the pack, but the effects of gravity and time should not be relied upon. Wireline devices will settle the pack.[19] After repair, the well should be logged again to confirm that the gravel pack has been settled successfully before it is placed on production.

Gravel-Pack Vibration

Techniques are available that are claimed to be capable of vibrating the gravel pack to create low pack porosity.[21] These methods impart mechanical energy to the screen or slotted liner during the packing process to rearrange the gravel grains. Gravel-pack vibration is believed to be highly desirable if it can be accomplished under field conditions. Field-scale laboratory tests have demonstrated that low-viscosity fluids pack at a porosity of about 39 to 40%.* Vibrating these 20-ft-long gravel packs reduced the porosity to about 35%. Tests with slurry-pack fluids indicated that they pack at a porosity of 41 to 42%. Pack vibration before the viscosity was broken indicated that pack settling was slow and that the porosity could not be reduced to a low level without breaking or removing the viscous fluid before vibrating.

Summary

1. Gravel placement involves transporting the gravel from the surface to the completion interval to install a downhole filter.
2. Gravel can be placed with water- or oil-based low- or high-viscosity fluids.
3. Gravel can be placed with either conventional or slurry placement techniques.
4. Fluid dehydration to a minimum gravel porosity yields the best gravel pack.
5. Mechanical techniques for placing gravel include wash-down, reverse-circulation, and crossover methods.
6. Blank sections in the screen assembly should be minimized when high-viscosity fluids are used to prevent voids in the gravel pack.
7. Deviated wells require special design considerations so that gravel can be placed effectively. This usually involves using large-diameter wash pipe and pumping at higher rates.
8. Logging techniques are available for determining whether voids exist in the gravel pack.
9. Gravel-pack vibration may lower pack porosity.

*Unpublished report, Exxon Production Research Co., Houston (1984).

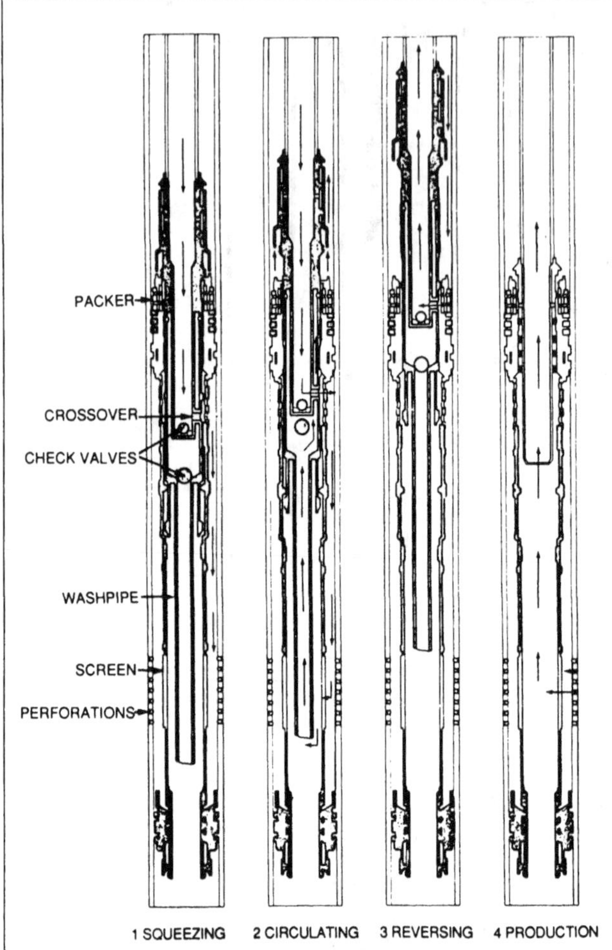

Fig. 8.19—Schematic of a crossover-type gravel pack with sump packer (cased hole).

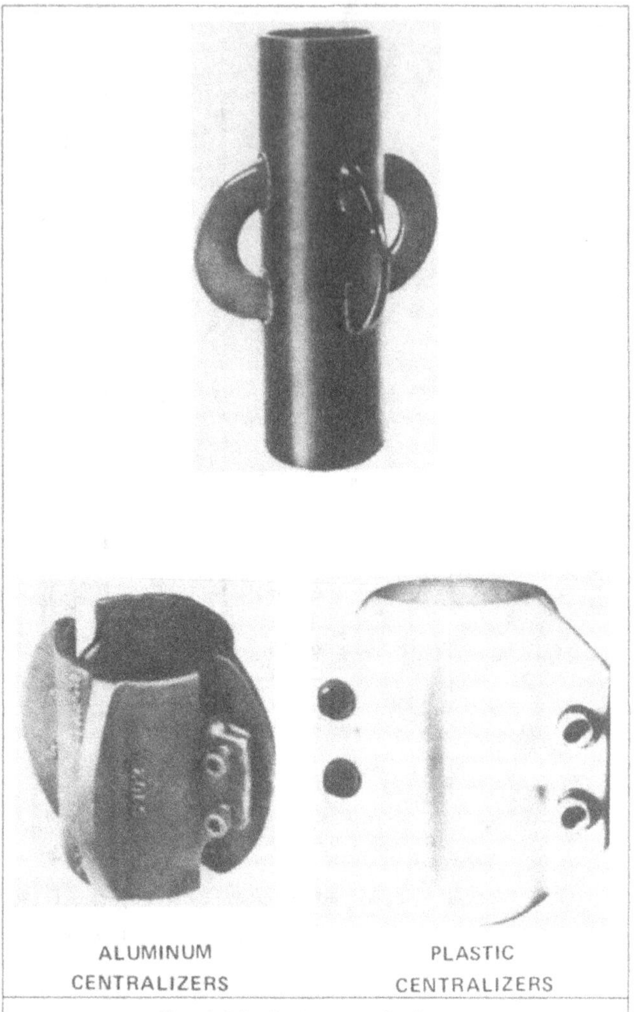

Fig. 8.20—Screen centralizers.

References

1. Elston, T.D. and Anderson, G.W.: "Foam Gravel Packing," paper SPE 11013 presented at the 1982 SPE Annual Technical Conference and Exhibition, New Orleans, Sept. 26–29.
2. Underdown, D.R., Das, K., and Nguyen, H.: "Gravel Packing Highly Deviated Wells With a Crosslinked Polymer System," *JPT* (Dec. 1985) 2197–2202.
3. Scheuerman, R.F.: "Guidelines for Using HEC Polymers for Viscosifying Solids-Free Completion and Workover Brines," *JPT* (Feb. 1983) 306–14.
4. Scheuerman, R.F.: "A New Look at Gravel-Pack Carrier Fluid," *SPEPE* (Jan. 1986) 9–16.
5. van Ballegooyen, J., Giap, T.K., and Seng, T.K.: "Experience in Gravel Packing Long, Perforated Intervals in Deviated Holes," *JPT* (Nov. 1983) 2079–86.
6. Torrest, R.S.: "The Flow of Viscous Polymer Solutions for Gravel Packing Through Porous Media," paper SPE 11010 presented at the 1982 SPE Annual Technical Conference and Exhibition, New Orleans, Sept. 26–29.
7. Torrest, R.S.: "Aspects of Slurry and Particle Settling and Placement for Viscous Gravel Packing (AQUAPAC)," paper SPE 11009 presented at the 1982 SPE Annual Technical Conference and Exhibition, New Orleans, Sept. 26–29.
8. Sparlin, D.D.: "Fight Sand with Sand—a Realistic Approach to Gravel Packing," paper SPE 2649 presented at the 1969 SPE Annual Meeting, Denver (Sept. 28–Oct. 1).
9. Suman, G.O. Jr., Ellis, R.C., and Snyder, R.E.: *Sand Control Handbook*, second edition, Gulf Publishing Co., Houston (1983).
10. Shryock, S.G., Dunlap, R.G., and Millhone, R.S.: "Preliminary Results From Full-Scale Gravel-Packing Studies," *JPT* (June 1979) 669–75.
11. Gruesbeck, C. and Collins, R.E.: "Particle Transport Through Perforations," *SPEJ* (Dec. 1982) 857–65.
12. Rensvold, R.E. and Decker, L.R.: "Full Scale Gravel Packing Model Studies," paper EUR39 presented at the 1978 SPE European Offshore Petroleum Conference, London, Oct. 24–27.
13. Shryock, S.G. and Millhone, R.S.: "Gravel-Packing Studies in a Full-Scale, Vertical Model Wellbore—Progress Report," *JPT* (July 1980) 1137–43.
14. Shryock, S.G.: "Gravel-Packing Studies in a Full-Scale, Deviated Model Wellbore," *JPT* (March 1983) 603–09.

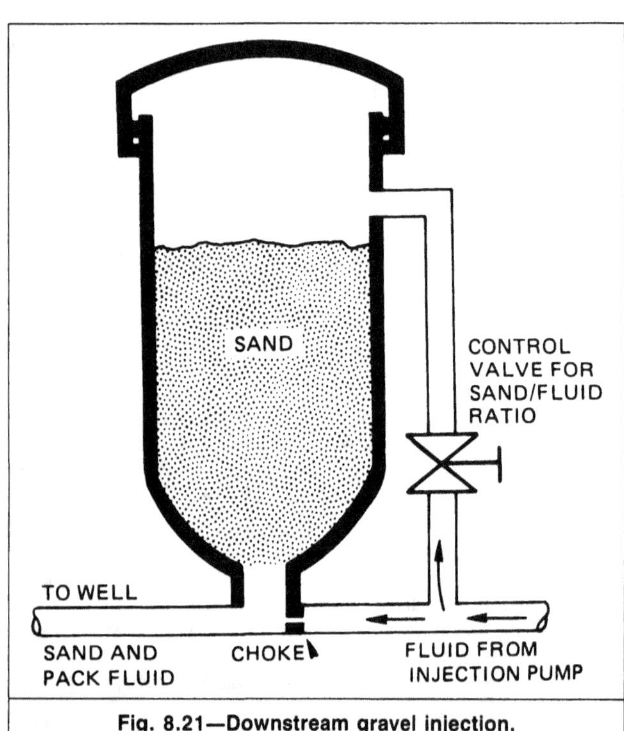

Fig. 8.21—Downstream gravel injection.

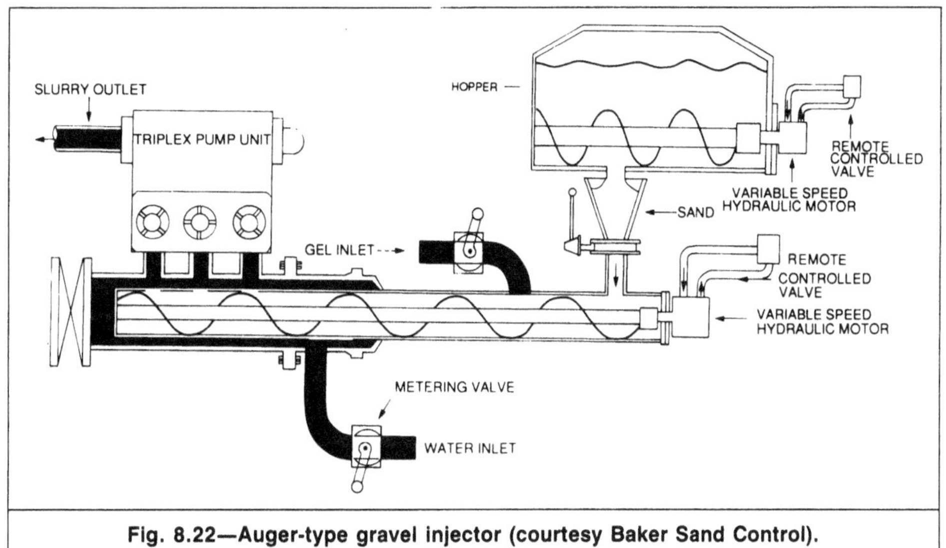

Fig. 8.22—Auger-type gravel injector (courtesy Baker Sand Control).

15. Penberthy, W.L. Jr. and Echols, E.E.: "Gravel Placement in Wells," paper SPE 22793 presented at the 1991 SPE Annual Technical Conference and Exhibition, Dallas, Oct. 6-9.
16. Maly, G.P., Robinson, J.P., and Laurie, A.M.: "New Gravel Pack Tool for Improving Pack Placement," *JPT* (Jan. 1974) 19-24.
17. Nini, C.J. and Owen, G.W.: "Successful High-Angle Gravel Packing Techniques," paper SPE 12105 presented at the 1983 SPE Annual Technical Conference and Exhibition, San Francisco, Oct. 5-8.
18. Schroeder, D.E. Jr. "Gravel Pack Studies in a Full-Scale, High-Pressure Wellbore Model," paper SPE 16890 presented at the 1987 SPE Annual Technical Conference and Exhibition, Dallas, Sept. 27-30.
19. Neal, M.R.: "A Quantitative Approach to Gravel Pack Evaluation," *JPT* (June 1985) 1035-40.
20. Carroll, J.F., Neal, M.R., and Parker, W.B.: "Gravel Pack Repair with Wireline Devices," paper SPE 15274 presented at the 1986 SPE Formation Damage Control Symposium, Lafayette, LA, Feb. 26-27.
21. Solum, J.R.: "A New Technique in Sand Control Using Liner Vibration With Gravel Packing," paper SPE 12479 presented at the 1984 SPE Formation Damage Symposium, Bakersfield, Feb. 13-14.

SI Metric Conversion Factors

bbl	× 1.589 873	E−01	=	m^3
cp	× 1.0*	E+00	=	mPa·s
ft	× 3.048*	E−01	=	m
°F	(°F−32)/1.8		=	°C
gal	× 3.785 412	E−03	=	m^3
in.	× 2.54*	E+00	=	cm
lbm	× 4.535 924	E−01	=	kg

*Conversion factor is exact.

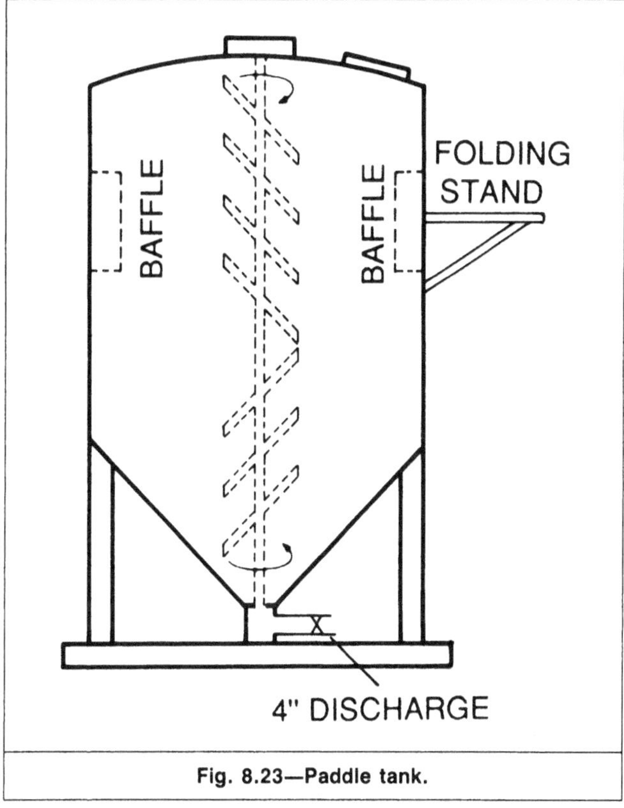

Fig. 8.23—Paddle tank.

Chapter 9
Screen Installation and Gravel-Pack Procedure Guidelines

Introduction

The techniques for installing screens and performing gravel packs fall into four categories (perforation cleaning, prepacking the perforations, conventional gravel packing, and crossover gravel packing), discussed in the procedures that follow. These general procedures are intended to serve as guidelines in situations where the engineer is not familiar with these operations. Individual circumstances and well conditions may require departures from the procedures outlined.

Perforation Cleaning

Wash Tool—Reverse-Circulation Type.

1. Make up the tool going in the hole, with the distance between the upper swab cups equal to the length of the perforated interval plus 10 ft.
2. With the lower swab cup in the bypass position, run the workstring or assembly into the hole slowly to 10 ft above the top perforation.
3. Close the bypass around the lower swab cup and test the tool to 500 psi by applying annulus pressure with the annular preventer closed.
4. With the annular preventer closed and 200-psi pressure applied down the annulus, lower the tool to the first perforation. Mark the workstring at the top perforation, indicated by a pressure loss or the start of circulation.
5. Begin reverse circulation with filtered fluid at the minimum rate required to transport sand up the workstring to the surface.
6. Wash the perforations with 10 bbl fluid/ft of interval or until pressures stabilize. Brine is the preferred wash fluid.
7. Continue washing until the bottom perforation is reached, which is signaled by an increase in pressure as the lower swab cup passes below the bottom perforation.
8. Release the bypass around the lower swab cup and reverse circulation until the wellbore is clean.
9. Repeat Steps 5 through 8 from the top of the perforations downward.
10. With the bypass open around the lower swab cup, perform a muleshoe prepack (see Prepacking the Perforations section).

Note: Perforation washing rates should be high enough to lift the wash debris from the well. A lift velocity of 100 ft/min usually is sufficient.

Wash Tool—Conventional-Circulation Type. First make up the wash tool and check the operation of the circulating valve. That is, the up position of tubing indicates that the valve is closed; the down position, that it is open. (Rotational-type valves are not recommended because left-hand torque is usually required to open the tool.)

Proceed with the perforation washing operation in the following sequence.

1. Run the workstring in the hole to below the perforations. Raise it slightly and test the tool to 500 psi down the workstring.
2. With 200 psi on the tubing, slowly raise the workstring until a loss of pressure indicates that the bottom perforation has been reached. Mark the workstring.
3. Initiate circulation down the workstring at 1 to 1.5 bbl/min using filtered fluid. Increase the pump rate as much as possible without causing fracturing.
4. Wash perforations with 10 bbl fluid/ft. Brine is the preferred wash fluid.
5. Slack off and open the circulating valve. Reverse circulation to clean the well.
6. Repeat Steps 4 and 5 until the entire perforated interval is cleaned.
7. Slowly pull the workstring from the hole (1,000 to 1,500 ft/hr), taking extreme care to maintain the hole full of fluid.

Note: Specific situations may dictate higher pump rates and larger wash-fluid volumes. Annular velocities should be about 100 ft/min for efficient removal of wash debris.

Surge Tool. When the surge-tool procedure is implemented, one should design the tool string with the following considerations in mind.

- Atmospheric-chamber collapse strength must be capable of withstanding full hydrostatic load plus 1,000-psi packer test pressure, including a safety factor of 1.25.
- Atmospheric-chamber volume should create a 1,000-psi differential pressure into the wellbore.
- Packer-setting depth should be such that the casing storage volume below the packer is greater than the atmospheric chamber volume.

Then make up the tool string in the following order.

1. Check valve (flow allowed, injection prevented).
2. Pup joint.
3. Mechanically set packer (must hold pressure from either direction and set with right-hand rotation).
4. Pup joint.
5. Surge valve (straight pull or left-hand rotation to open).

6. Atmospheric chamber with length designed according to considerations listed above.

7. Landing nipple with plug (preferably wireline-retrievable) installed.

8. One joint of tubing.

9. Ported collar or sliding sleeve.

10. Tubing, as required, to space out the packer to calculated setting depth. (Test above and below the atmospheric chamber to 5,000 psi. Torque turn or externally test the atmospheric-chamber tubing.)

Then take the following steps.

1. Enter the hole to the required packer-setting depth. Fill the tubing every two or three stands.

2. Set the packer and test the annulus to 1,000 psi.

3. Connect the kill manifold to the tubing annulus and test the lines to working pressure.

4. Open the surge valve as dictated by the type of valve used (rotation or pull).

5. Rig up the wireline unit, equalize, and pull the plug from the landing nipple above the chamber.

6. If pressure exists on the tubing, try to bleed it off. Do not bleed more than 2 or 3 bbl. If the pressure does not bleed down, open the ported collar and reverse the tubing clean of produced fluids. Use the fluid weight, as required, to control surface pressure.

7. Release the packer and reverse circulate through the kill manifold and choke.

8. When clean fluids return and the well is completely under control, lower the pipe and reverse circulate to clean the well to below perforations.

9. Slowly pull out of the hole.

Prepacking the Perforations

Muleshoe Method.

1. Enter the hole with open-ended tubing to 5 ft below perforations. Close annular preventers and reverse circulate with filtered fluids.

2. Test injectivity at 1 to 2 bbl/min and record pressure on the annulus. If the pressure level is high (and the rate is low), perform a mud-acid treatment.

3. Initiate gravel circulation at a concentration of ¾ to 1 lbm/gal in filtered fluid.

4. As the sand slurry nears the end of the tubing, close the annulus return line and begin injection.

5. Alternately move the pipe up 5 ft and down 4 ft for about 1 minute. Continue reciprocating the pipe until the entire perforated interval is packed and the wellbore is full of sand to above the perforations, which is indicated by a high injection pressure.

6. Slowly reverse the wellbore clean to below the perforations and estimate the volume of sand placed outside them.

7. Repeat Steps 3 through 6 until no additional sand is placed outside the perforations.

8. Reverse circulate the well around at least two tubing volumes.

9. Pick the workstring up above the perforations and shut the well in for 30 minutes.

10. Lower the pipe and tag the bottom. If more than 1 ft^3 of gravel has fallen back into the wellbore, repeat prepack steps 3 through 9. If the wellbore is clean, pull out of the hole. If gravel continues to fall into the wellbore, spot a high-viscosity pill across the perforated interval.

11. Slowly pull out of the hole to prevent swabbing prepack gravel into the wellbore.

Conventional Gravel-Pack Methods

Wash-Down Gravel-Pack Method. This procedure should not be used if the screen setting is longer than the working height of the rig.

1. Clean the perforations with one of the washing-tool or surge-tool procedures described earlier.

2. Prepack the perforations, following Steps 1 through 7 of the previous section.

3. After completing the prepack until gravel is no longer being placed outside the perforations, leave gravel inside the wellbore to a height calculated by

$$h_g = \frac{C_{an}L_s}{C_c},$$

where

h_g = height to gravel left in casing,
C_{an} = capacity (vol/ft) of annulus between screen OD and casing ID,
L_s = length of screen and blank liner setting, and
C_c = capacity (vol/ft) of casing ID.

4. Make up the screen setting in this sequence:
- Shoe (standard or turbojet).
- Screen with centralizers.
- Blank liner with centralizers.
- Back-off sub.

5. Hang the screen off in slips and make up the wash pipe.

6. Space out the wash pipe using a slip-joint releasing tool so that the pipe seals into the set shoe.

7. Make up the tubing to go in the hole with the screen setting to the top of the gravel fill.

8. Initiate circulation down the tubing at 1 to 2 bbl/min.

9. Slowly lower the screen to the required completion depth while circulating the cleaning fluid.

10. Stop the circulation and wait 15 minutes.

11. Apply annulus pressure and check for complete gravel coverage of the screen. (If required, reverse gravel down the annulus to fill in around the screen, according to the procedure described in the Reverse Circulation section.)

12. With 5,000-lbf pull above the pipe weight, rotate the tubing to the right and release it from the back-off sub.

13. Lift the screen 3 to 5 ft and clean the well with reverse circulation.

14. Pull out of the hole with the tubing.

15. Run a tieback overshot with a mechanically set packer to the top of the screen.

16. Locate and space out so that the overshot will seal onto the top of the screen setting. Test the annulus at the final packing pressure and watch for circulation, indicating that the overshot is not sealed.

17. Set the packer and test the annulus to 1,500 psi.

18. Release the packer or nipple up the tree, as production requirements dictate.

Reverse-Circulation Method. First clean the perforations according to one of the wash-tool or surge-tool methods described earlier, followed by the muleshoe prepack method. The gravel pack procedure is normally performed as follows.

1. Make up the screen setting as follows:
- Bladed bull plug.
- Screen with centralizers.
- Blank liner with centralizers.
- Telltale screen joint, with a 2-ft screen located 10 to 15 ft below the top of the setting.
- Back-off sub.

2. Set the screen setting in slips and make up the wash pipe and releasing tool. Space the wash pipe to within 1 ft of the bottom of the screen and make up the releasing tool to the back-off sub.

3. Enter the hole to the required setting depth.

4. Initiate reverse circulation with a gravel slurry at 1 lbm/gal in filtered fluid.

5. After filling the annulus to above the screen, calculate the volume required to fill the annulus to the telltale.

6. Reverse circulate 75 to 80% of the calculated gravel volume and follow that with clean fluid.

7. If the telltale is not covered, reverse circulate a 1- or 2-bbl gravel slurry followed by clean fluid.

8. Repeat Step 7 until the telltale is covered, as noted by a large pressure drop. Do not exceed the blank-liner collapse rating.

9. Release the workstring from the screen and reverse the well clean. Pull out of the hole.

10. Run a tieback overshot with a mechanical packer and space out so that the overshot is over the back-off sub.

11. Test the tieback seal at the final pack pressure and watch for circulation.

12. Set the packer and test to 1,500 psi.

13. Release the packer or nipple up the tree, as dictated by production requirements.

Crossover Gravel-Pack Methods

Swab-Cup Crossover Tool. Before implementing the swab-cup crossover technique, first clean and prepack the perforations. Make up the screen with the following components:

- Bladed bull plug.
- Screen with centralizers.
- Blank liner with centralizers.
- Telltale screen joint, with a 2-ft screen section 10 to 15 ft below the top of joint.
- Back-off sub.
- Crossover tool.

Then take the following steps.

1. Slowly run the assembly to the required depth.
2. Begin circulating filtered fluid down the tubing.
3. Circulate down the tubing a slurry composed of 1 lbm gravel/gal fluid.
4. After the first pressure increase (which will occur when gravel fills above the screen section), circulate enough additional slurry to fill 75% of the blank-liner section above the screen.
5. Follow the slurry volume with clean fluid.
6. Circulate 1 bbl of slurry mixed at 1 lbm gravel/gal fluid and displace the substance with clean fluid.
7. Repeat Step 6 until the telltale section is covered, signaled by a large pressure increase. Do not exceed the blank-liner collapse rating.
8. Release from the top of the screen assembly and raise the workstring 2 to 3 ft.
9. Clean the well by reverse circulation; then pull out of the hole.
10. Run a tieback assembly and packer.
11. Locate the tieback assembly over the back-off sub and test to the final pack pressure.
12. Set the packer and test the annulus.
13. Release the packer or nipple up the tree, as dictated by production requirements.

Packer/Crossover-Tool Combination

Single Completion (Circulation Pack). In the single-completion packer/crossover technique, clean and prepack the perforations as previously discussed. Then make up the screen assembly with these components:

- Bladed bull plug.
- Screen with centralizers.
- Blank liner with centralizers.
- Telltale screen joint, with a 2-ft screen wrapped 10 ft below the top of the joint.
- Safety joint (straight-pull release pinned at 80% tubing strength minus the tubing weight).
- Pup joint.
- Port-collar/packer assembly.

Then perform the following steps.

1. Hang the screen assembly in slips and make up the wash pipe, crossover tool, and packer.
2. Make up the packer to screen assembly. (The packer and crossover tool should be tested before makeup).
3. Enter the hole to the required depth and space out tubing so that a 5-ft stroke can be achieved without stripping a collar through the annular preventer.
4. Set the packer according to the manufacturer's recommendations and test the annulus for packer seal.
5. Manipulate the crossover tool to the circulation position and initiate circulation down the tubing.
6. Circulate the gravel slurry at 1 lbm/gal of filtered fluid.
7. As the slurry nears the bottom, close the circulating port on the crossover tool and bullhead the first 10 to 15 bbl of slurry into the perforations.
8. Open the crossover tool to the circulating position to continue circulating the slurry around the screen until a pressure increase is noted. This increase indicates complete coverage of the screen with gravel. (A pressure increase at this point may be very small, depending on the pump rate, screen size, and other conditions.)
9. Calculate the volume of gravel required to cover the blank liner and circulate 75% of that volume; follow it with clean fluid.
10. Circulate 1 bbl of slurry with a 1 lbm/gal gravel concentration, followed by clean fluid.
11. Repeat Step 10 until a large pressure increase is noted, indicating coverage of the telltale-screen section.
12. Release the crossover tool from the packer and raise the workstring to the reversing-out position.
13. Reverse circulate to clean the wellbore and pull out of the hole.
14. Run production tubing as required.

Dual Completion. In the dual-completion packer/crossover technique, follow the above procedures for single completions through Step 13.

1. Set a permanent packer 10 ft below the upper-zone perforations.
2. Run a packer plug with a fishing neck and set in the permanent packer. (If the lower-zone packer is within 10 to 15 ft of the upper zone, making up the screen assembly can be eliminated and the plug set in the lower-zone packer.)
3. Perforate the upper zone, clean the perforations, and prepack according to the procedures outlined earlier. A full-opening plug-retrieving tool can be run on the end of the workstring for the prepack operation, thereby eliminating a trip-out later to pick up the tool.
4. Spot a high-viscosity pill of 50 to 100 cp across the upper perforations.
5. Retrieve the packer plug and pull out of the hole.
6. Run the upper-zone screen and crossover assembly as follows:
   - Seal assembly and locator sub for permanent packer.
   - Landing nipple with standing valve installed.
   - Sealbore receptacle (smaller ID than the screen or upper-zone crossover packer).
   - Screen.
   - Blank pipe.
   - Telltale joint, with 2 ft of screen wrapped 10 ft below the top of the joint.
   - Safety joint.
   - Pup joint.
   - Port collar.
   - Pup joint as required to space out the crossover tool.
7. Hang the screen assembly in slips and make up the wash pipe with the seal assembly, slip-joint crossover tool, and packer. Space out so that the seal assembly is less than 6 in. into the sealbore receptacle.
8. Make up the packer/screen assembly. (The packer and crossover tool should be tested before makeup.)
9. Enter the hole to the permanent packer and space out the seals.
10. With the seal assembly just above the permanent packers, initiate circulation at 1 to 2 bbl/min or at the minimum rate required to maintain circulation.
11. Lower the seals into the packer and check for loss of circulation to test seal engagement.
12. Set the gravel-pack packer and test the annulus.
13. With the crossover tool in the circulating position, start the gravel slurry down the tubing.
14. As the slurry nears the end of the tubing, close the crossover tool and bullhead the slurry into the formation.
15. When a pressure increase is noted, indicating coverage of the screen with gravel, open the crossover tool. Circulate a gravel slurry volume equal to 75% of the calculated volume required to fill the annulus around the blank liner. Follow it with clean fluid.
16. Circulate 1 bbl of slurry followed by clean fluid.

17. Repeat Step 16 until a large pressure increase is achieved, which indicates coverage of the telltale screen by gravel.

18. Release from the packer and pick up to the reverse-circulating position with the crossover tool.

19. Reverse the well clean and pull out of the hole.

20. Pull the standing valve from the nipple in the upper screen assembly.

21. Run dual-completion equipment as required.

Single Completion (Slurry Pack). In the single-completion packer/crossover technique, clean and prepack the perforations. Then make up the screen assembly with the following components:

- Bladed bull plug.
- Telltale screen joint.
- Blank joint (length will depend on the size of the rathole interval).
- Screen with centralizers.
- Blank joint (length will depend on the amount of gravel reserve desired).
- Safety joint (straight-pull release pinned at 80 to 90% tubing strength minus tubing weight).
- Pup joint.
- Port-collar/packer assembly.

Follow these steps.

1. Hang the screen assembly in slips and make up the wash pipe, crossover tool, and packer.

2. Make up the packer/screen assembly. (The packer and crossover tool should be tested before makeup.)

3. Enter the hole to the required depth and space out tubing so that a 5-ft stroke can be achieved without stripping a collar through the annular blowout preventer.

4. Set the packer according to the manufacturer's recommendations and test the annulus for packer seal.

5. Pump into the formation to check injectivity. If it is low, mud-acidize the interval.

6. Manipulate the crossover tool to the circulation position and initiate circulation down the tubing at 3 to 5 bbl/min.

7. Circulate gravel slurry (10 to 15 lbm/gal of filtered fluid) down to the crossover.

8. Just before the slurry reaches bottom, reduce the rate to 0.5 bbl/min.

9. Close the circulating port on the crossover tool and squeeze the slurry into the perforations until the limiting pressure level is reached.

10. Open the crossover tool to the circulating position and attempt to circulate slurry around the screen until the limiting pressure is reached.

11. Return to the squeeze position and repeat Step 8.

12. Release the crossover tool from the packer and pick up to the reversing-out position.

13. Reverse circulate the wellbore clean and pull out of the hole.

14. Run production tubing as required.

SI Metric Conversion Factors

bbl	×	1.589 873	E−01	=	m^3
cp	×	1.0*	E+00	=	mPa·s
ft	×	3.048*	E−01	=	m
ft^3	×	2.831 685	E−02	=	m^3
gal	×	3.785 412	E−03	=	m^3
in.	×	2.54*	E+00	=	cm
lbf	×	4.448 222	E+00	=	N
lbm	×	4.535 924	E−01	=	kg
psi	×	6.894 757	E+00	=	kPa

*Conversion factor is exact.

Chapter 10
Gravel-Pack Performance

Introduction

Probably the best evaluation of any completion technique is its field performance. One must remember, however, that field performance involves human, operational, reservoir, and other factors that are not necessarily a part of completion technology but affect gravel-pack results. After a sufficient number of completions, however, trends in performance can be established. These trends are beneficial in planning procedures for future wells and projecting performance on the basis of experience. Unfortunately, field experiences are not well documented and are often taken out of context. Consequently, good examples of field performance are not always available. Engineers do not always take the time to document results when they occur and later the details become misstated, lost, or confused.

Studies of gravel-pack performance* have considered many of the factors discussed in previous chapters. Some of the items studied were the effects of gravel-pack type (open or cased hole), prepacking, perforating, and type of fluid production (oil, water, or gas) on completion success. Also, new completions were compared with workovers. The following discussion reviews the effects of these variables on gravel-pack completion productivity and lifetime. Remember that particular operational, reservoir, and other conditions may be site-specific and cause results that deviate slightly from the trends discussed here.

Completion Type

Openhole Gravel Packs. Openhole gravel packs have the highest productivity of all gravel-pack completions, primarily because of the large inlet area surrounding the screen at the sandface that reduces the resistance to flow. Another important factor is that, most of the time, the completion interval is underreamed to allow a thicker gravel layer to be placed around the screen that may extend completion longevity. Underreaming also reduces the flow resistance in the immediate vicinity of the well. The combination of these parameters accounts for the higher PI's experienced with such completions.

Table 3.1 lists the PI's for various gravel-packed completions in two different reservoirs. The PI's are quite different in the two reservoirs, but the relative similarity in the PI behavior of these completions shows that the openhole gravel pack has at least twice the productivity of other perforated gravel-packed completions. Although these data were taken from Miocene reservoirs in Venezuela during the early 1960's, the experience gained relates to gravel-packed wells throughout the world. In this particular example, the perforating technology consisted primarily of smaller-diameter (0.4-in.) perforations as opposed to the large-diameter guns currently available. Current technology has reduced the discrepancy between open- and cased-hole completion productivity.

The performance of openhole gravel packs is unique in that, if wells can be completed successfully, they produce satisfactorily for extended periods of time. That is, any failure is observed initially rather than at some later date. This is shown in **Figs. 10.1 and 10.2,** which compare openhole and cased-hole gravel-pack performance. This longevity holds for oil and gas injection and for water-source wells. Several of the water-source-well openhole gravel packs in Fig. 10.2 have actually produced more than 25 million bbl. These examples represent total equivalent production. For gas wells and associated gas, equivalent liquid production was calculated at a ratio of 1 bbl=8 Mscf. See Chap. 2 for a definition of success.

Cased-Hole Gravel Packs. Cased-hole gravel packs are used more widely than openhole gravel packs because of drilling, completion, and workover considerations. These completions generally do not have the productivity experienced with openhole gravel packs owing to flow-through gravel- or formation-sand-filled perforations. Properly designed perforation and completion programs, however, have substantially increased well productivity.

Unlike openhole gravel packs, cased-hole gravel packs generally show a decrease in completion success as a function of time or cumulative production. This result is believed to occur because perforations become plugged with formation material, causing restricted productivity and completion failure. Also, the cumulative production from cased-hole gravel packs is usually not as high as that for openhole gravel packs. Fig. 10.2 shows both of these observations.

As discussed previously, prepacking gravel in the perforations increases completion life and well productivity, as the examples in **Figs. 10.3 and 10.4** illustrate. The effect of the increased flow rate shown in Fig. 10.3 supports previous arguments concerning the desirability of prepacking. Because completion longevity is increased (Fig. 10.4), the time interval between workovers lengthened. Each of these factors has a positive effect on cash flow.

Cased-hole gravel packs are amenable to almost any casing size. From a practical standpoint, the larger sizes are more attractive because they will accommodate standard-sized tools and equipment. But cased-hole gravel packs have also been used in 2⅞- and 3½-in. tubingless completions with some success. Tubingless completions are hampered by the use of slim-hole tools and equipment and by perforating programs that will accommodate only guns that shoot small-diameter charges. **Fig. 10.5** demonstrates experience with the tubingless-completion gravel packs. These packs have good initial success but high decline rates. The decline probably reflects

*Unpublished reports, Exxon Production Research Co., Houston (1976).

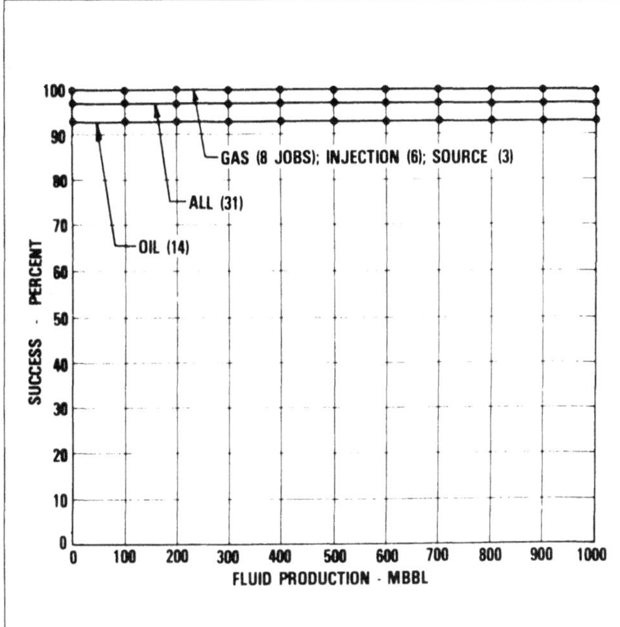

Fig. 10.1—Effectiveness of openhole gravel packs by well type (courtesy Exxon Production Research Co.).

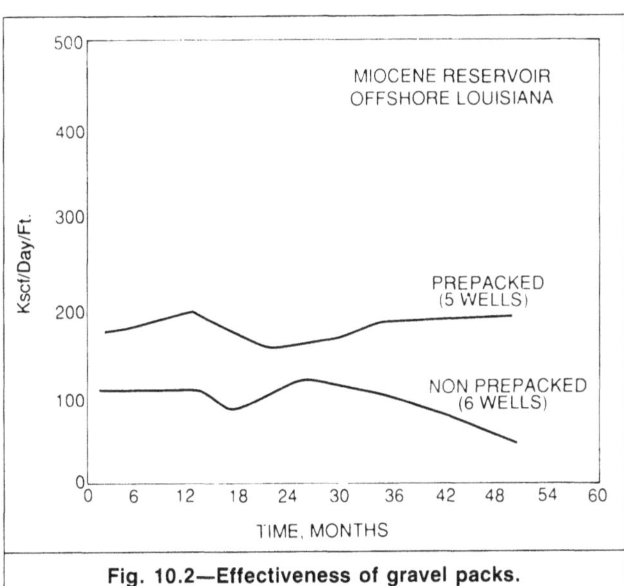

Fig. 10.2—Effectiveness of gravel packs.

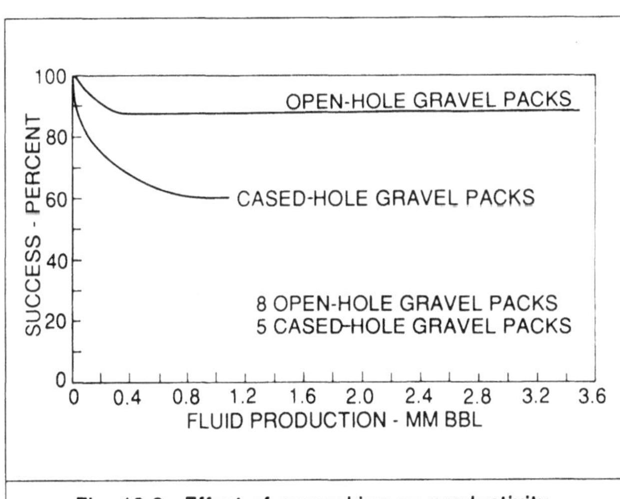

Fig. 10.3—Effect of prepacking on productivity.

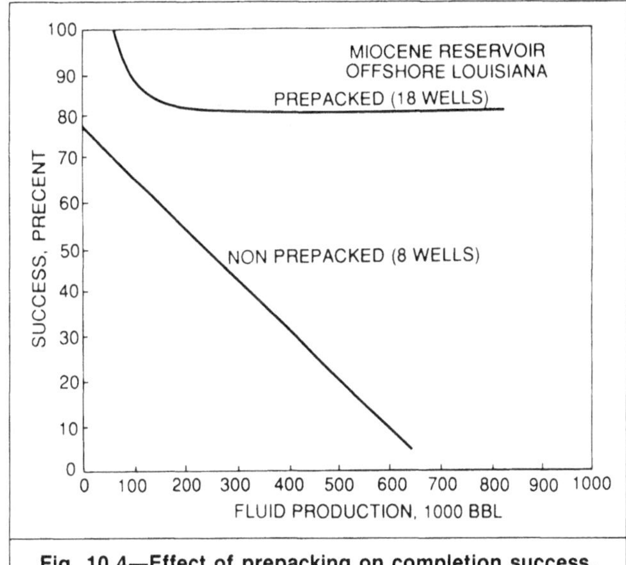

Fig. 10.4—Effect of prepacking on completion success.

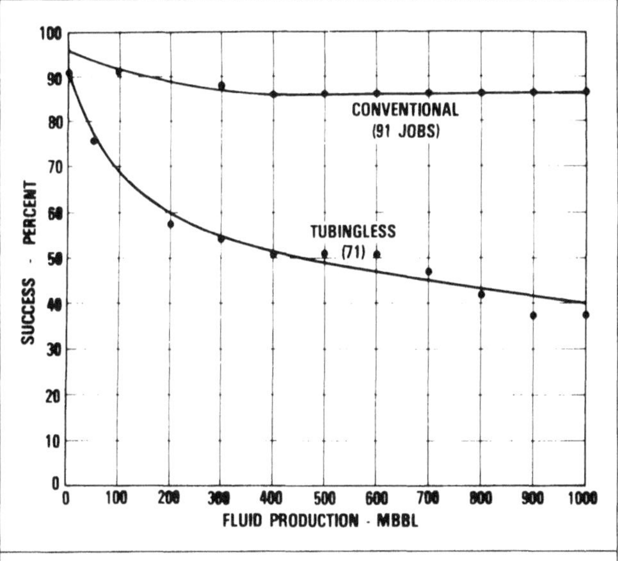

Fig. 10.5—Effect of wellbore configuration on gravel packs.

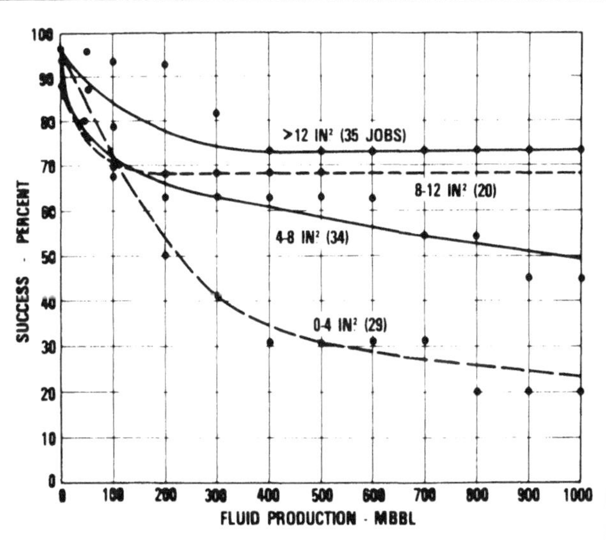

Fig. 10.6—Effectiveness of cased-hole gravel packs relative to perforated flow area.

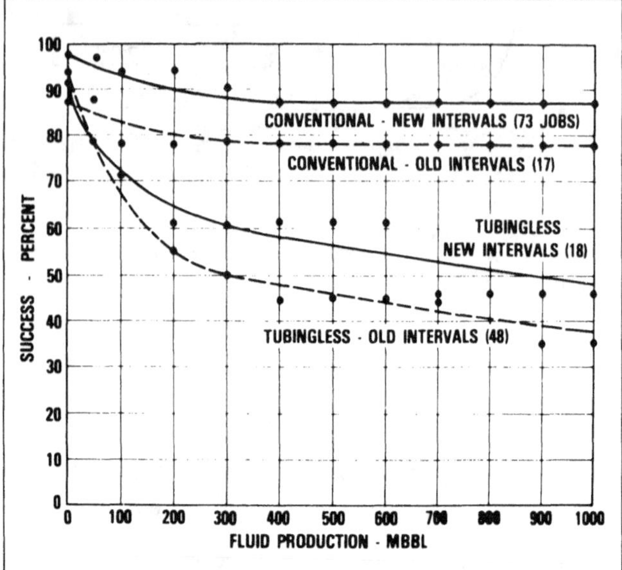

Fig. 10.7—Effect of interval age and wellbore configuration on gravel packs.

the influence of a limited number of small-diameter perforations becoming plugged with formation sand and restricting fluid inflow because perforation cleaning operations were not conducted on these tubingless wells. In this example, a conventional completion has a 4½-in.-diameter or larger casing.

Perforation size affects cased-hole gravel-pack success, but the total perforation area is also important. This point is illustrated in **Fig. 10.6,** which shows that completion success is higher with large perforation areas than with small perforation areas. This particular example does not differentiate between large- and small-diameter perforations. The implication here is that, with high-inlet areas, the lower fluid velocities through the perforations tend to reduce the drag forces on the formation. The result is higher cumulative production for the high-perforation-area wells. The tendency for perforations to plug is less than in cases where lower perforation densities tend to increase the fluid velocity in the perforation and have a greater tendency to restrict flow and to plug.

Another comparison is that between cased-hole gravel packs in new and old intervals (**Fig. 10.7**). A "new interval" means it is the initial gravel pack on that interval; in an "old interval," the interval is being recompleted. Field results for these situations show that new intervals are more successful than older intervals. Whether this result is controlled by the effectiveness of workover operations in general or affected by other factors is not clear. However, regardless of whether a conventional or tubingless completion was gravel-packed, the new intervals show about a 10% increase in success compared with old intervals.

Other observations regarding gravel-pack performance are as follows.*

1. The success rate of gravel packing is not appreciably affected by wellbore deviations for angles up to 50°.

2. Acidized gravel packs correlate with high failure rates, especially in wells treated for low productivity after they had been produced for an extended period of time. However, wells acidized upon initial completion usually respond satisfactorily to acidizing. One of the problems with hydrofluoric acid treatments is gravel dissolution. It can also be a problem with steam injection if bicarbonate ions are present in the boiler feedwater and the stream condensate has a pH of 9.5 or higher.[1] Steam injection can also fluidize the gravel pack or pump it through perforations so that, when wells are returned to production in steam stimulation operations, a gravel pack no longer exists. Hence, sand control is lacking.

3. The technique used to place the gravel does not appear to be a factor in success or failure as long as the proper gravel and screen size and the correct procedures are used to complete the well.

These performance histories suggest that many variables can influence gravel-pack longevity. As a general statement, a small percentage of all gravel-packed completions produce more than 1 million bbl of fluid. One reason is that reserves of this magnitude may not be under the influence of the well's drainage. Obviously, many favorable conditions must exist for the well to produce this volume of fluid. However, some wells have produced substantially more than 1 million bbl of fluid. Wells with this potential are usually completed in clean, thick, high-permeability reservoirs.

Summary

- Openhole gravel packs have greater longevity than cased-hole gravel packs.
- Prepacking enhances productivity and longevity.
- Gravel packs in conventionally completed wells show greater longevity than those in tubingless completions.
- High-perforation areas enhance completion longevity.

Reference

1. Reed, M.G.: "Gravel Pack and Formation Sandstone Dissolution During Steam Injection," *JPT* (June 1980) 941–49.

SI Metric Conversion Factors

ft^3	× 2.831 685	E−02	=	m^3
in.	× 2.54*	E+00	=	cm

*Conversion factor is exact.

*Unpublished reports, Exxon Production Research Co., Houston (1976).

Chapter 11
Plastic Consolidation Principles

Introduction

Plastic consolidation is a method of stopping sand production by artificially bonding the formation sand grains into a consolidated mass. A liquid resin is pumped through the perforations and into the pore spaces of the formation sand. The resin coats the sand grains and then hardens to hold the sand in place.

Consolidation Objectives

To control sand successfully, several important objectives must be reached during consolidation. **Fig. 11.1** is a schematic of these objectives. The primary objective of a consolidation treatment is to bond together the formation sand adjacent to every perforation.

Because the treatment is designed to function on formation sand, a zone several feet in radius around the wellbore can be consolidated. Attaining sufficient strength to withstand the fluid pressure gradient during production requires that three objectives be achieved at the microscopic level: coat the sand grains with resin, concentrate the resin at the contact points to bond the grains together, and leave the bulk of the pore space open for flow. A photograph from a scanning electron microscope (**Fig. 11.2**) illustrates the achievement of these objectives in an actual consolidation of formation sand.

The purpose of coating the sand grains with resin is to wet the exposed sand surface, thereby forming a good adhesive bond. Concentration of the wetting phase (resin) at the grain contact points is a natural occurrence in all porous media when two phases are present. Interfacial tension (IFT) between the resin and the liquid filling the remainder of the pore space causes the resin (wetting phase) to be drawn into the grain contact regions. Before the resin hardens, the pore space is partially displaced with a nonreactive fluid. Oil is usually used for this purpose, but water may be used with some systems. Approximately 35% PV is filled with resin, leaving the remainder open for flow. This amount will vary among sand consolidation systems and according to the permeability and porosity of the formation sand.

Fig. 11.3 shows a laboratory wellbore model after consolidation. In the laboratory test, a 1-in. pipe with two 0.5-in. perforations spaced 1 ft apart was surrounded with formation-type sand. Consolidating fluid was pumped through the pipe and perforations into the sand. It can be seen that the consolidation extends a full 360° around the well; the consolidated radius is about 1 ft. Fig. 11.3 shows two distinct lobes of consolidated sand, which seem to meet at a point halfway between the perforations. Depending on the vertical permeability, the consolidating fluids can flow vertically to encase the entire wellbore. Had there been a plugged perforation between the two open ones, sand adjacent to the plugged perforation would have been treated by consolidating fluids pumped through the open perforations.

In practice, many formations contain shale breaks that limit the vertical flow of consolidating fluids to fairly short distances. Hence, it is important to inject the consolidation chemicals through as many perforations as possible. The necessity of consolidating the sand adjacent to every perforation cannot be overemphasized. The fact that a perforation does not accept injected fluid during consolidation is no guarantee that it will not permit fluid (and sand) flow during production.

The question of how large a consolidated radius is necessary for a successful treatment often arises. Ideally, a consolidated zone a few inches thick around the wellbore is all that is required. Because strata of different permeabilities accept injected fluids in different ratios, and because not all perforations are open, it is impossible to achieve the ideal consolidation in the field. Increasing the radius of the consolidated sand mass can help cover plugged perforations that did not accept resin injection. In addition, all consolidation treatments lose some strength with time, which increases the consolidated radius needed to withstand the drawdown stress imposed by fluid production. Experience has shown that an average consolidated radius of 2 to 3 ft is a good compromise between observed treatment success and costs.

Consolidation Chemicals

Sand consolidation treatments use a variety of chemicals, either alone or in combination. These chemicals can be grouped into five functional categories: preflushes, resins, diluents, coupling agents, and overflush fluids.

Preflushes. Surface preparation is vital in any consolidation operation. Resin coating (wetting) of the sand grains must be sufficient to provide adhesive strength. Preflushes are used ahead of the resin to control the pore-fluid composition and sand-surface characteristics so that an adequate resin coating is achieved. Preflushes consist of oil, water, or mutual solvents.[1-7]

Water or oil preflushes are used to remove undesirable formation brine or crude from the pore spaces near the wellbore. Natural formation fluids frequently contain constituents that are incompatible with the consolidation resin. Because crude oils can dissolve or absorb a portion of the resin, the cured strength will be lower than anticipated. Similarly, the salinity of the formation brine may be too high or too low for optimal resin performance.

Adding a surfactant to an oil or water preflush will reduce IFT and attain a more complete removal of natural formation fluids.[2] In addition, the wettability of the formation sand can be modified with surfactants. The effect of a surfactant on the sand wettability generally is much less predictable than its effect on the IFT. For this reason, preflush surfactants are carefully formulated to meet

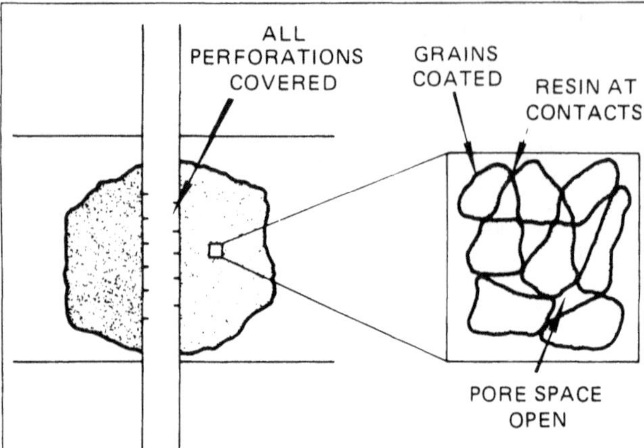

Fig. 11.1—Plastic-consolidation objectives: (a) macroscopic; (b) microscropic.

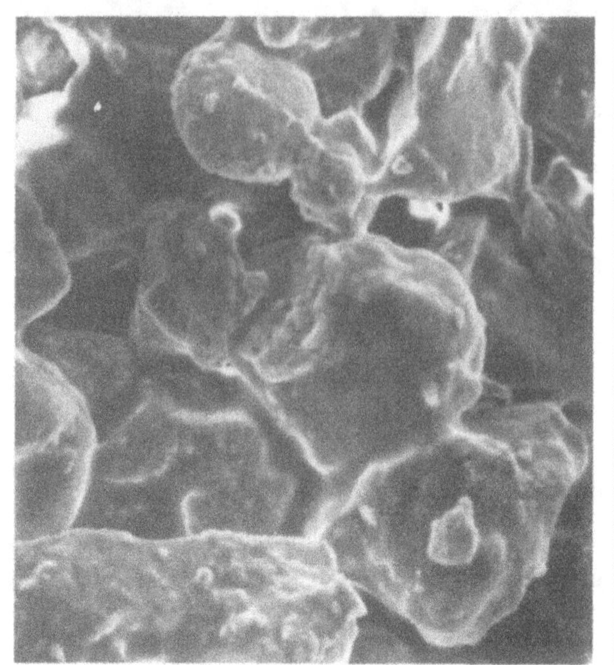

Fig. 11.2—Plastic distribution within a pore space as seen by a scanning electron microscope.

the requirements of each specific consolidation resin.

The term "mutual solvent" refers to a fluid that can dissolve substantial amounts of brine and oil simultaneously. A mutual-solvent preflush can remove almost all the residual water and oil from the pore space. Mutual solvents can be preferentially water miscible or preferentially oil miscible. If a preferentially water-miscible mutual solvent is added to equal volumes of brine and crude, the volume of the aqueous phase will increase but the oil phase will show little change in volume. As **Fig. 11.4a** shows, the addition of more solvent eventually dissolves both the oil and aqueous phases. This type of mutual solvency is most effective in mobilizing water in the formation pore space. A preferentially oil-miscible mutual solvent will primarily enlarge the oil phase, as shown in **Fig. 11.4b**, and is most effective in mobilizing oil.

Choosing the type of mutual-solvent preflush presupposes knowledge of which fluid (water or oil) should be preferentially removed. Laboratory experience has shown that most resins perform better if water is removed from the pore space.[2] This is not surprising because, in most sands, water is the wetting phase that must be displaced before the resin can coat the sand. Glycol ethers are widely used as mutual solvents. The monoglycol ethers tend to be preferentially oil miscible while the polyglycol ethers usually are preferentially water miscible. Other common mutual solvents include acetone and isopropyl alcohol.

Fig. 11.3—Plastic distribution around a wellbore.

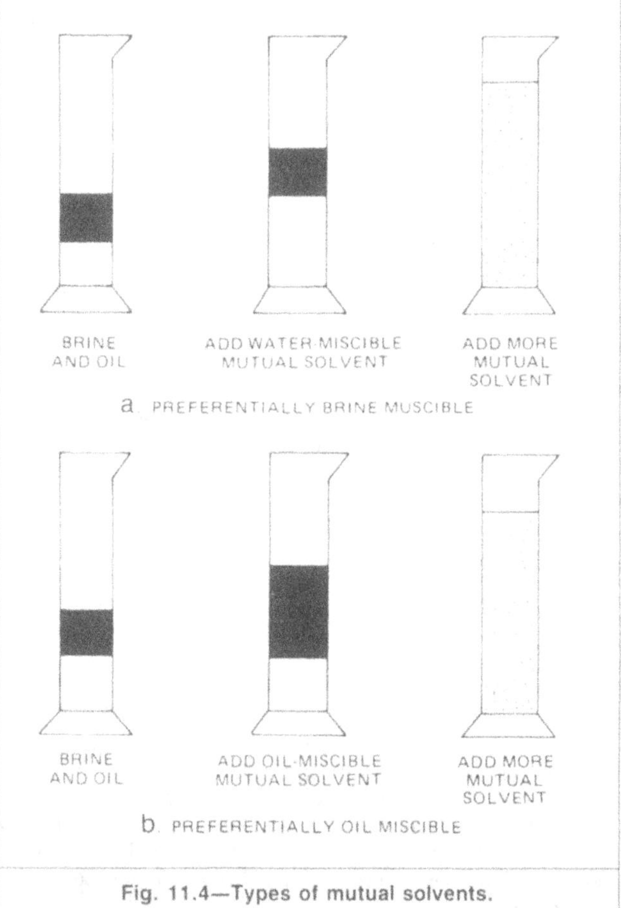

Fig. 11.4—Types of mutual solvents.

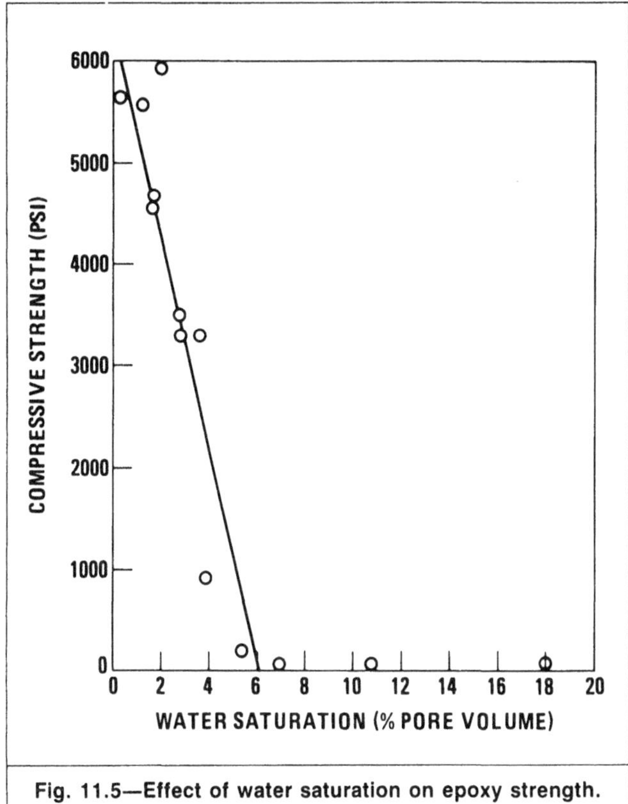

Fig. 11.5—Effect of water saturation on epoxy strength.

-C-C(-O-) + C(-O-)-C- ⟶ -C-C(OH)-O-C-C- (Catalyzed)

-C-C(-O-) + NH_2- ⟶ -C-C(OH)-NH- (Amine Cured)

a. **Epoxy**

furan-CH_2OH + furan-CH_2OH ⟶ furan-CH_2-furan-CH_2OH + H_2O

b. **Furan**

phenol(OH) + CH_2O + phenol(OH) ⟶ phenol(OH)-CH_2-phenol(OH) + H_2O

C. **Phenolic**

Fig. 11.6—Consolidation chemistry.

TABLE 11.1—SUMMARY OF RESIN PROPERTIES

Resin	Catalyzed or Cured by	Reaction Byproducts	Softening Temperature (°F)	Wets Sand in Presence of Water
Epoxy	Amines	None	250	No
Furan	Strong acid	H_2O	300	Yes
Phenolic	Strong base	H_2O	300	Yes

In determining the amount of mutual solvent required as a preflush, it is necessary to define how the consolidation strength is affected by residual water saturation. **Fig. 11.5** shows this relationship for an epoxy resin system and reveals that sufficient mutual solvent must be injected ahead of the resin to reduce the water saturation to less than 3 to 4% PV.[7] Otherwise, the consolidation strength will be too low to be effective.

Resins. Although many commercial plastic consolidation processes are available, most are based on the polymerization reactions of an epoxy, a furan, or a phenolic resin. **Table 11.1** summarizes the basic reaction chemistry and physical properties of these resins.

Resin polymerization proceeds through either catalysis or curing. During catalyzed polymerization, a small amount of catalyst causes the resin molecules to join together into long chains. Curing differs from catalyzing in that the curing agent reacts directly with the resin and becomes an integral part of the resin structure. Despite these differences, the terms "catalyzed" and "cured" are frequently used interchangeably.

Epoxy. Epoxy resins are well known for their adhesive and surface-coating capabilities. The polymerized resin contains highly polarized sites that give it great affinity for polar surfaces such as silica. As **Fig. 11.6a** shows, epoxy curing reactions produce no reaction byproducts. Also, there is little shrinkage during cure. These factors combine to produce epoxy coatings that are smooth, uniform, and highly resistant to chemical attack. All consolidation resins lose strength as the temperature increases, and epoxy resins are no exception. Above 250°F, ordinary epoxy formulations become soft and yield when placed under stress. This upper temperature limit does not significantly restrict usage for sand control because most unconsolidated sands are found at temperatures below 250°F.

Epoxy resins are sensitive to the amount of residual water in the sand.[6,7,9,10] Plain epoxies will not displace water from a silica surface, and bonding does not occur when there is a layer of water between the resin and the sand. For this reason, a water-removing preflush or surfactant must be used with an epoxy resin.

Furan. Furan consolidation systems are based on the polymerization of furfuryl alcohol by an acid catalyst. The furfuryl alcohol molecules join together by forming a methylene bridge ($-CH_2-$) and releasing water as a byproduct. **Fig. 11.6b** shows this process. As Table 11.1 shows, the catalyst is usually a strong acid, such as trichloroacetic acid (oil soluble) or hydrochloric acid (water soluble). Ordinarily, a mixture of furfuryl alcohol and furan resin is used instead of plain alcohol. The furan resin is prepared by partially polymerizing some furfuryl alcohol and then removing the produced water. Furan polymerization reactions occur very quickly. Some spontaneous polymerization occurs even at room temperature, which can limit the shelf life of some resins.

Furan resins exhibit good chemical resistance to all common chemicals, except such oxidizing agents as chlorine. The water released during polymerization tends to collect as tiny bubbles in the plastic coating that can act as stress concentration sites. Furan resins are able to coat the sand grains in the presence of water, so a mutual-solvent preflush for water removal is not necessary. The upper temperature limit for these resins is about 300°F, which is slightly higher than for epoxies.

Phenolic. Phenolic resins were used in some of the earliest consolidation processes.[5] Phenolic resin is formed by the reaction between a phenol and an aldehyde. As **Fig. 11.6c** shows, the result is a methylene bridge linking two phenol molecules. Because each phenol molecule has many sites to which the aldehyde may attach, the final structure is highly crosslinked. Phenolic reactions may be promoted with either an acid or a base catalyst; however, a base such as sodium hydroxide is generally preferred.

Table 11.1 shows that the physical properties of phenolic resins are quite similar to those of furans. Phenolics have good chemical resistance and a softening temperature of 300°F. The water produced as a reaction byproduct creates defects in the coating that can lead to cracking and stress concentration.

Some resin formulations use a mixture of phenolic and furan chemistry. Phenolic resins are diluted with furfuryl alcohol and then polymerized with an acid catalyst. The result is a hybrid structure of phenol and furan rings linked with methylene bridges.

Diluents. Almost every resin suitable for sand consolidation is too viscous to be pumped in the undiluted state. Therefore, diluting the resin becomes an important consideration in formulating a sand consolidation process. Starting with undiluted resins with viscosity of 100 to 10,000 cp, diluent levels of 10 to 50 wt% are commonly required to reach the desired 10- to 20-cp diluted viscosity. Two types of diluents, reactive and nonreactive, are available for resin dilution. A reactive diluent is able to polymerize with the resin.

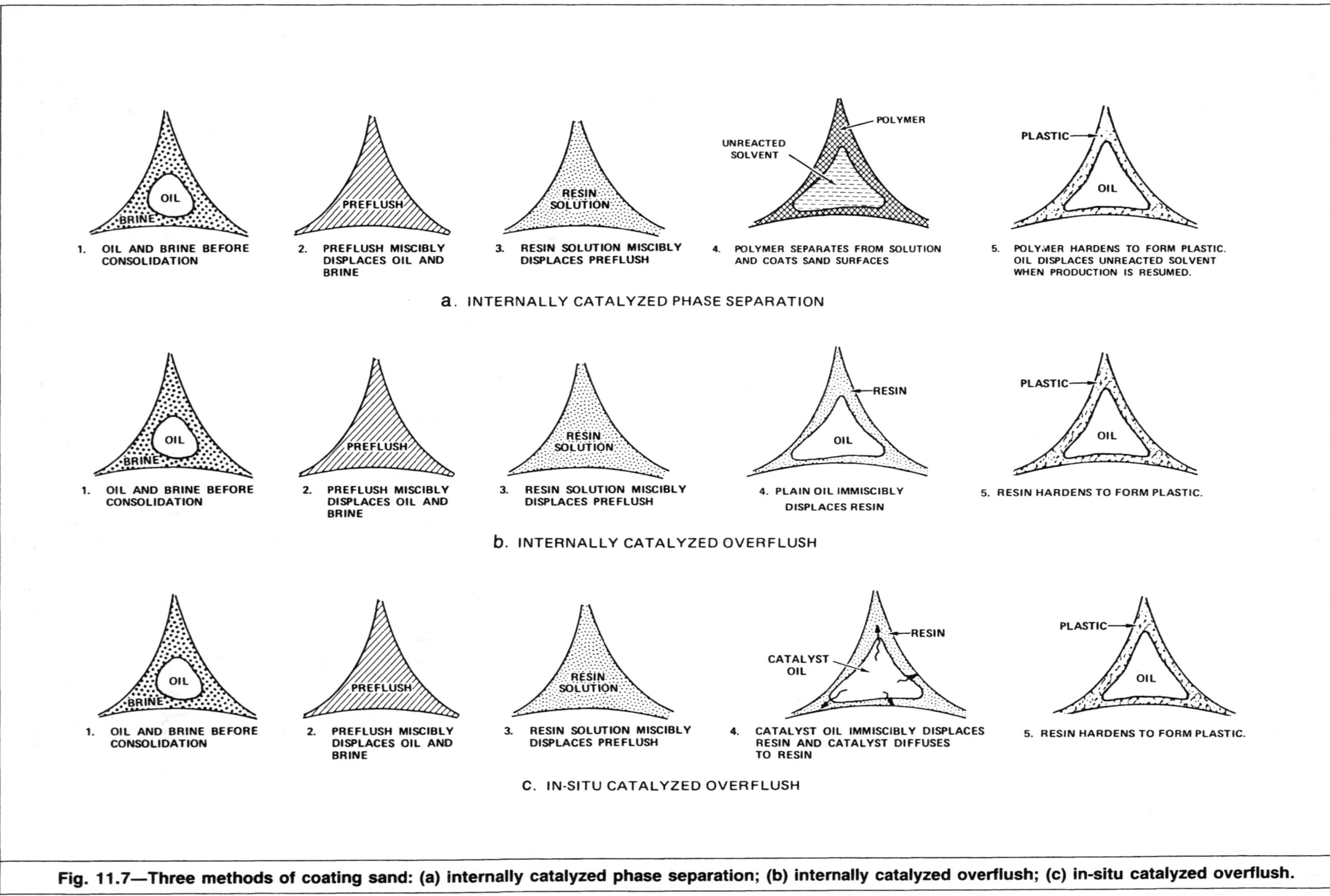

Fig. 11.7—Three methods of coating sand: (a) internally catalyzed phase separation; (b) internally catalyzed overflush; (c) in-situ catalyzed overflush.

Fig. 11.8—Example of poor plastic coating caused by incomplete wetting: (a) low magnification; (b) high magnification.

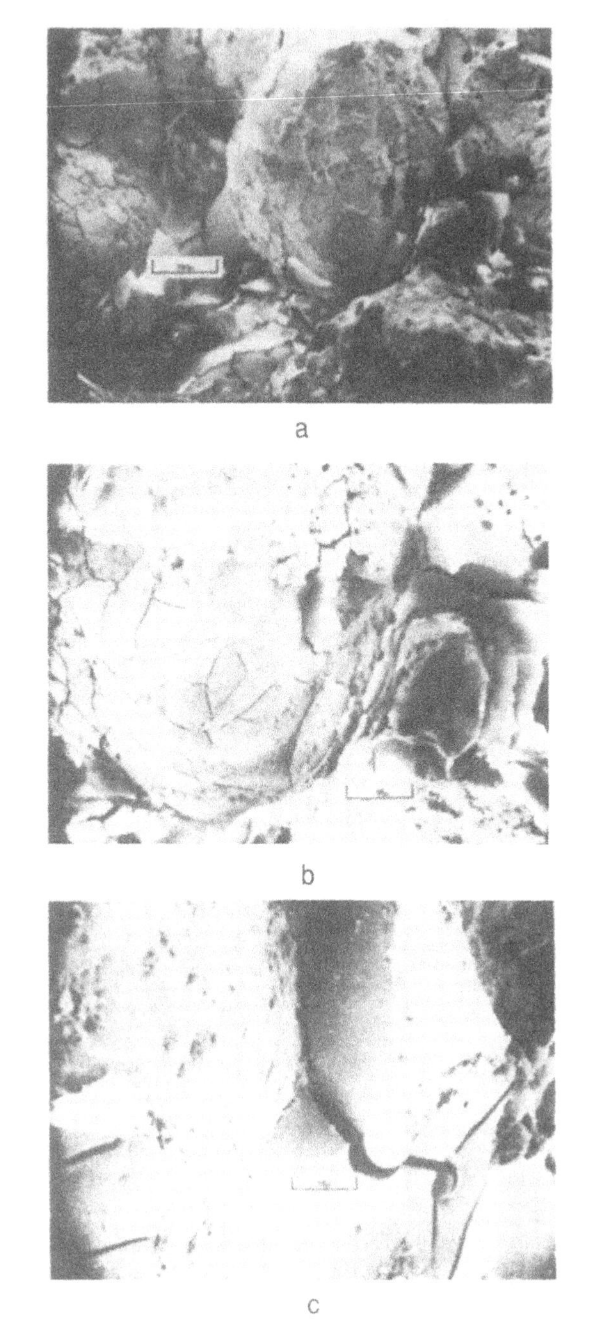

Fig. 11.9—Example of surface cracking owing to shrinkage: (a) low magnification; (b) medium magnification; (c) high magnification.

A nonreactive diluent does not become part of the resin structure and is frequently removed during the polymerization process.

Coupling Agents. Coupling agents are used in a plastic consolidation process specifically to bond the resin to the silica surface chemically. These chemical bonds are much stronger than the usual attractive forces between resin and sand. The most common coupling agents are organosilanes. The organic end group (such as an amine) reacts with the resin during polymerization; the silane end group reacts with the sand.

$$H_2NCH_2CH_2NH-(CH^2)_3Si(OCH_3)_3$$

reacts with resin — reacts with silica

The coupling agent is often included as an ingredient in the resin formulation, but some studies have shown that coupling agents are more effective when added to the preflush.

Overflush Fluids. An overflush fluid is often injected after the plastic consolidation resin. This overflush performs three important functions. First, it opens a fluid pathway through the resin so that the consolidated sand will be permeable. Second, it often must carry the catalyst into the pore spaces to harden the resin coating. Third, the overflushing fluid is sometimes required to extract the diluent from the resin. To displace the excess resin efficiently from the pore space, the overflush fluid should be more viscous than the resin. This ensures a favorable mobility ratio during displacement, which will prevent the fluid from fingering or channeling through the resin.

For resins that require an oil overflush, price and availability make diesel the most widely used oil. However, diesel has two drawbacks. First, diesel viscosity is only 3 to 5 cp as opposed to 10 to 20 cp for the typical resin. This difference makes the mobility ratio unfavorable for displacing resin with diesel. Second, diesel exhibits appreciable solvency for some consolidation resins. Part of the resin is lost owing to dissolution in the diesel.

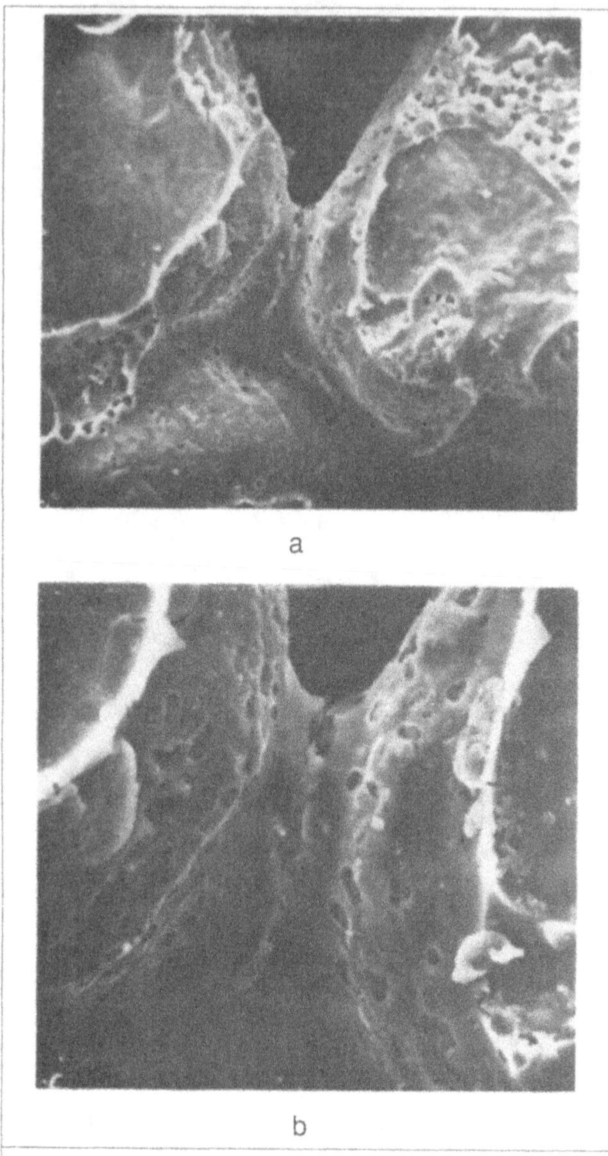

Fig. 11.10—Example of bubble defects in plastic coating: (a) low magnification; (b) medium magnification.

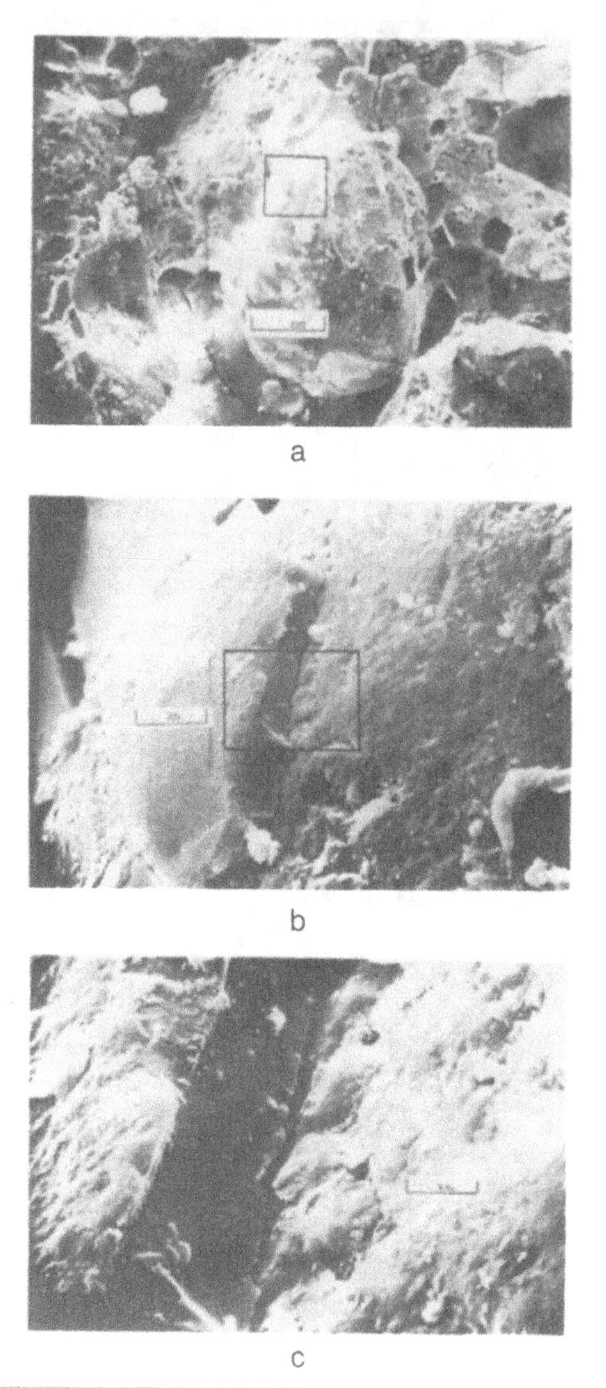

Fig. 11.11—Example of adhesive failure: (a) low magnification; (b) medium magnification; (c) high magnification.

Several systems use oils that are more highly refined than diesel. These include motor-oil bright stock, white oil, and process oil. These more viscous oil grades make controlling the overflush/resin mobility ratio easier. Resin solubility is also reduced owing to the more paraffinic nature of these refined oils.

Water (brine) can be used as an overflush fluid with some resin systems. It is more economical than diesel, but the mobility-ratio problem is compounded because water's viscosity is even less than that of diesel.

Sand-Coating Methods

A consolidating resin should harden as a thin coating on the sand grains but leave the bulk of the pore space open for fluid flow. Although many commercial consolidation processes are available, there are only three basic means of achieving the desired resin coating: internally catalyzed phase separation, internally catalyzed overflush, and in-situ catalyzed overflush.

Fig. 11.7 and the following sections describe these processes without referring to the specific type of resin used. Applications to specific resins are described in Chap. 12. The name given to each coating method refers first to the mechanism of mixing the catalyst with the resin and second to the means of opening the pore space for fluid flow.

Internally Catalyzed Phase Separation. The internally catalyzed phase-separation method involves direct mixing of the resin components and the catalyst before they contact the sand (hence, the description "internally catalyzed"). "Phase separation" refers to the precipitation of the polymerized plastic from the initial resin solution. At the end of the polymerization, two phases exist in the pore space: polymerized plastic, which coats the sand, and the solvent from the original resin solution, which fills the bulk of the pore space and provides permeability.[5]

Fig. 11.7a is a step-by-step representation of the pore-space fluid saturations for an internally catalyzed phase-separation plastic. Initially, the pore space is filled with brine and oil, which are then displaced with a preflush. The resin solution then displaces the preflush to fill the pore space. At this point, fluid injection ceases and the curing period begins. As the resin polymerizes, it becomes insoluble in the carrier solvent and precipitates (phase separates) onto the sand grains. This leaves the carrier solvent in the center of the pore space and a hardened resin film on the sand.

Mixing the resin and catalyst before injection gives precise control over the resin/catalyst ratio. Because the resin begins to polymerize as soon as the catalyst is added, the resin solution must be injected promptly into the formation before the polymerization proceeds sufficiently to cause phase separation. Ordinarily, there is enough time between catalyst addition and phase separation for pumping, but there would not be sufficient time in case of a serious mechanical breakdown.

Internally Catalyzed Overflush. All the resin ingredients used in the internally catalyzed overflush technique—including the catalyst (or hardener)—are blended before injection. Permeability is achieved by immiscible displacement of the excess resin by an inert fluid (usually an oil).[4] Fig. 11.7b is a schematic of this process. The oil and brine are displaced by preflush, which in turn is displaced by resin. At this point, an oil is used to displace excess resin immiscibly and to establish permeability. The sand grains are left coated with resin, which then hardens to cement the grains together. The overflush process may be likened to flushing a water-saturated sand to residual water with oil. With the consolidation process, a "residual resin" saturation is established by the overflushing procedure. Typical residual resin saturations are 30 to 35% PV.

In-Situ Catalyzed Overflush. Unlike the techniques described above, the catalyst in the in-situ catalyzed overflush procedure is not included in the resin solution. Instead, it is dissolved in the overflushing fluid and diffuses into the resin phase once the pore space has been overflushed.[3,6,7] Fig. 11.7c is a schematic of the process. Separating the resin and catalyst during plastic placement overcomes the risks of the limited pumping time associated with internally catalyzed processes. In-situ catalyzed resins retain their original fluid state until they come in contact with an overflush fluid containing catalyst.

Consolidation Failure Mechanisms

The mechanism of plastic deterioration involves both cohesive failure of the plastic, as evidenced by the development of a large number of cracks in the coating, and subsequent adhesive failure of the bond between the plastic and the grain surface.[11,12] All commonly available plastic consolidation systems seem to undergo this type of deterioration. Some types of plastics deteriorate much more rapidly, however, because their inherent physical and chemical properties accelerate the deterioration process. In general, four failure mechanisms have been identified: incomplete wetting, shrinkage cracks, bubble defects, and adhesion failures.

Incomplete Wetting. Complete grain coating may not be achieved because of poor wetting by resin. The two photomicrographs in **Fig. 11.8** illustrate this effect. The figure shows several sand grains that are only partially coated with plastic. Note that the plastic formed small balls on the surface of the sand grains. This occurrence suggests that as the resin cured to form a plastic, it acted as a nonwetting liquid in contact with the grain. One likely cause of this behavior is ineffective water removal by the preflush.

Shrinkage Cracks. Fig. 11.9 illustrates cracks attributed to plastic shrinkage during polymerization. The figure shows a low-magnification view of a field sample recovered from a well after failure of an initially successful treatment. At higher magnification, an extensive and uniform network of cracks is clearly visible in the plastic coating. Because the sand grains are fairly rigid, plastic coatings that undergo significant shrinkage during polymerization will develop high internal stresses that can lead to cracking.

Bubble Defects. Cracking can also be initiated by stress concentrations at defect sites. **Fig. 11.10** shows the junction of three sand grains bonded by plastic. Many bubbles are apparent in the plastic coating, and cracks in the bonding region between the sand grains are clearly visible. The bubbles are sites of stress concentration from which a network of cracks can easily initiate and propagate throughout the bonding region. Bubble defects may be caused by water produced as a byproduct of the polymerization reaction. As the resin cures, its molecular weight and viscosity increase. Because the produced water is insoluble in high-molecular-weight resin, it phase-separates to form droplets. If these droplets are trapped by the viscous resin, they can create bubble defects.

Adhesive Failure. Failure of a consolidation treatment is ultimately the result of cohesive failure in the plastic coating, followed by adhesive failure at the plastic/sand interface. **Fig. 11.11** shows one of many examples of adhesive failure observed in laboratory samples. A single, initially-well-coated grain is prominent in the center of the photomicrograph. The upper portion of this grain has a bare surface exposed as a result of adhesive failure (Fig. 11.11a). Higher magnification of this region (Fig. 11.11b) shows the bare sand-grain surface on the right and the remaining plastic coating on the left. At high magnification (Fig. 11.11c), it is readily apparent that the plastic and sand have separated in adhesive failure. The contours of the plastic and the sand at the interface match perfectly, suggesting that the liquid resin wet the surface of the sand grain completely, then failed after the resin had cured to a hard plastic.

Summary

The objective of any sand consolidation treatment is to bond the formation together for a radius of 2 to 3 ft from the well without significantly restricting the permeability. The consolidation material typically is a plastic—an epoxy, a furan, or a phenolic resin. Regardless of the plastic used, the placement process includes preflushing, coating the sand grains with resin, and curing to form a hard sand mass.

Studies of consolidation failure mechanisms show that plastics used for sand consolidation (1) must be inert to reservoir fluids, (2) should shrink very little during curing, (3) must have strong wetting and adhesive properties, (4) should produce minimal reaction byproducts and should prevent occlusion of any byproducts in the cured plastic, and (5) should exhibit sufficient flexibility (toughness) that the plastic can deform without failure to relieve localized stress concentrations.

References

1. Brooks, F.A.: "EGMBE, An Improved Organic Preflush for Sand Consolidation Plastics," paper SPE 4776 presented at the 1974 SPE Formation Damage Control Symposium, New Orleans, Feb. 7–8.
2. Penberthy, W.L. Jr. *et al.*: "Sand Consolidation Preflush Dynamics," *JPT* (June 1978) 845–50.
3. Treadway, B.R., Brandt, H., and Parker, P.H. Jr.: "Studies of a New Process to Consolidate Oil Sands With Plastics," *JPT* (Dec. 1966) 1537–43.
4. Richardson, E.A. and Hamby, T.W.: "Consolidation of Silty Sands with an Epoxy Resin Overflush Process," *JPT* (Sept. 1970) 1103–08.
5. Brooks, F.A.: "Consolidation of Dirty Sands by Phenol-Formaldehyde Plastic," *JPT* (Aug. 1971) 934–38.
6. Brooks, F.A. *et al.*: "Externally Catalyzed Epoxy for Sand Control," *JPT* (June 1974) 589–94.
7. Shaughnessy, C.M., Salathiel, W.M., and Penberthy, W.L. Jr.: "A New, Low-Viscosity, Epoxy Sand-Consolidation Process," *JPT* (Dec. 1978) 1805–12; *Trans.*, AIME, **265.**
8. Murphey, J.R., Bila, V.J., and Totty, K.: "Sand Consolidation Systems Placed With Water," paper SPE 5031 presented at the 1974 SPE Annual Meeting, Houston, Oct. 6–9.
9. Bezemer, C. and Meijs, F.H.: "Development and Field Application of the Process for Sand Consolidation Employing EPOSAND® Resin," paper SPE 1591 presented at the 1966 SPE Annual Meeting, Dallas, Oct. 2–5.
10. Hamby, T.W. and Richardson, E.A.: "New Sand Control Process Works Well in Dirty Sands," *World Oil* (June 1968) 81–84.
11. Muecke, T.W.: "Factors Influencing the Deterioration of Plastic Sand Consolidation Treatments," *JPT* (Feb. 1974) 157–66.
12. Rensvold, R.F.: "Sand Consolidation Resins—Their Stability in Hot Brine," *SPEJ* (April 1983) 238–48.

SI Metric Conversion Factors

cp	× 1.0*	E+00	= mPa·s
ft	× 3.048*	E−01	= m
°F	(°F−32)/1.8		= °C
in.	× 2.54*	E+00	= cm

*Conversion factor is exact.

Chapter 12
Commercial Plastic Consolidation Systems

Introduction

Through the years, many plastic consolidation processes have been developed and commercialized. This chapter outlines the criteria useful in the evaluation of consolidation processes and describes the performance of several commercial systems on the basis of these criteria. Typical treatment designs for commercially available consolidation processes are also presented.

Evaluation Criteria

Selecting the best plastic consolidation process requires a comparison of strength, permeability, aging characteristics, and field results of the available systems. Other considerations, such as cost and method of application, are also important. Comments on certain evaluation parameters are presented below as aids in the use of the data presented in this section.

The term "initial properties" refers to the strength and permeability of consolidated sand immediately after the curing period has elapsed. In general, the consolidated sample has not been exposed to either brine or crude oil.

Compressive Strength. Uniaxial compressive strength is used in the laboratory to measure how well sand grains are cemented together.[1-8] In a producing well, however, the load on the consolidation comes in the form of a *drawdown stress* imposed by fluid flowing through the sand. Flowing fluid causes the consolidation to fail in a shear mode (tension plus compression).

To relate compressive strength to drawdown stress at failure, a series of laboratory experiments was conducted. Fluid (oil) was flowed through a series of consolidated samples of known compressive strength until they failed. The results showed that each 1,000 psi in compressive strength corresponds to 1,700 psi in drawdown stress.

Consolidation processes that yield compressive strengths of 1,000 to 2,000 psi can withstand drawdown forces of 2,000 to 3,000 psi—considerably above the drawdown pressure drop encountered in most wells. Thus, any plastic system capable of yielding 1,000 to 2,000 psi in initial compressive strength is fully capable of controlling sand initially.

Permeability Retention. Evaluating the permeability retention of a consolidation process involves measuring the absolute (single-phase) permeability of the sand before and after treatment.[1-8] This permeability ratio (after/before consolidation) is defined as *permeability retention.*

Permeability retention will be less than 100% because a portion of the original pore space has been filled with plastic. To develop significant strength in formation-type sand, about 35% of the pore space must be filled with plastic. Permeability retention of about 75% is the best that can be achieved without sacrificing compressive strength.

Permeability retention and, to some extent, compressive strength depend on the grain size of the sand used in testing. Large pore spaces are less affected by partial filling with plastic than small pore spaces. Typically, a plastic with 50% permeability retention on a 1-darcy sand will show 80 or 90% retention on a 10-darcy sand. Permeability-retention values measured on 10-darcy sand might not represent expected field performance. No industry-standard sand size exists for permeability-retention tests.

Remember that permeability retention does not correlate directly with productivity loss. The reason is that only a portion of the fluid pressure drop occurs in the 2- to 3-ft consolidated radius around the well. A consolidation process that has 70% permeability retention typically will decrease "potential" well productivity by only 15 to 20%. If the production rate is being restricted because the well produces sand, as is often the case, then a successful consolidation treatment will usually enhance the production rate.

Aging Characteristics. Almost any commercial consolidation process generating 1,000 to 2,000 psi in initial compressive strength would be acceptable if the properties did not change with time. Unfortunately, consolidation plastics deteriorate when exposed to reservoir fluids for an extended time. When strength drops below that necessary to withstand the drawdown forces, the plastic fails and the well begins to produce sand. Knowledge of the plastic's aging characteristics is important in evaluation of any consolidation process.[9,10]

Water (brine) is the reservoir fluid with the most influence on a plastic's lifetime. It apparently invades the bonding region between the plastic and sand, causing the adhesive bond to fail. Efforts to simulate this process under controlled laboratory conditions have had only limited success. The important variables that correlate with aging are time and drawdown stress; volume flowed (PV) is influenced strongly by consolidation permeability and may have less correlation with aging. The relative aging characteristics of several plastics have been determined, but predicting field life expectancy from laboratory data alone has not been possible.

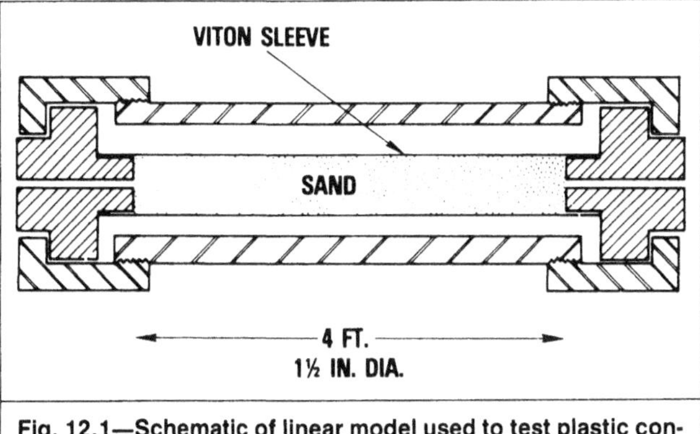

Fig. 12.1—Schematic of linear model used to test plastic consolidation characteristics.[4]

One surprising fact emerging from the laboratory aging data is that a significant strength loss can occur after just a few days of exposure to well fluids. This early, rapid loss is then followed by a more gradual decline; years often pass before failure occurs.

Because of the loss of strength with time, plastic systems with high initial strength can be expected to exhibit superior aging performance. Choosing a system on the basis of initial strength alone can be misleading, however, if information regarding the aging properties of that particular resin formula has not been considered.

Field Results. Actual field results are the most valuable data used to evaluate a plastic consolidation process. Because conditions vary so widely from well to well, however, a statistical approach must be used to evaluate job success. To be valid, this approach requires that a large number of treatments be performed. It is obviously impractical to generate field lifetime results for every consolidation process. But for some systems used extensively in the field, it is possible to correlate field results with properties revealed in laboratory tests.

Quite often, the success of a consolidation process in the field depends heavily on the complexity and reliability of the field procedure used to place the resin in the formation. This factor is not adequately reflected in compressive-strength and permeability-retention data. The placement factor further complicates the prediction of field performance from laboratory data. It can only be said that the low-strength systems may not be effective in the field and that the field performance of high-strength resins depends primarily on the technique used to place them.

Even actual field results are not guaranteed to predict future consolidation performance. The lifetime of a resin system depends heavily on how the wells are produced. Such variables as producing rate and water cut can significantly affect field results. These variables are seldom constant; they usually change with the degree of reservoir depletion.

Sources of Evaluation Data

Many data have been generated to help evaluate plastic consolidation systems. The information included in this section has been gathered from three sources: supplier and trade journal publications, laboratory data, and field experience.

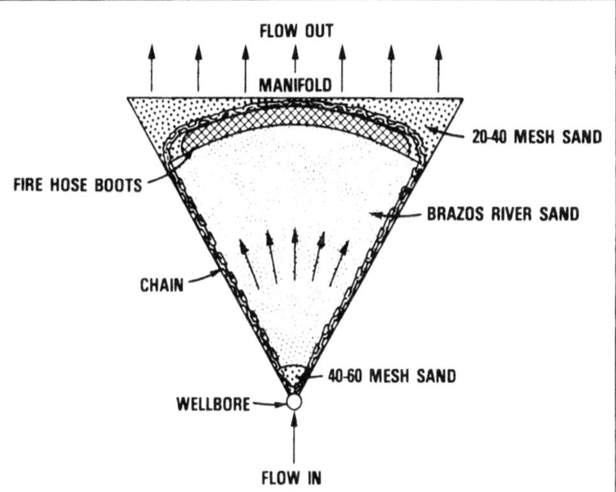

Fig. 12.2—Schematic of radial model used to test plastic consolidation under radial flow.[4]

Publications. Publications on commercial plastic consolidation systems are always available in the form of service company brochures or trade journal articles. Such descriptions invariably contain data on a system's performance under laboratory conditions. Because of the wide variation in testing methods, it is often difficult to compare data from several sources.

Laboratory Data. Most of the commercial consolidation processes have been tested in a single laboratory. Through the years, the basic variables of temperature, sand type, flow rate, and curing time have remained constant. This feature provides a fairly reliable basis for comparing the various processes.

Considerable laboratory testing has been performed with a 4-ft linear model (**Fig. 12.1**). Unconsolidated sand is packed in a 4-ft-long rubber sleeve 1.5 in. in diameter. Conditions in the sand are held at 1,000 psi and 140 to 160°F while a net overburden of 500 psi is applied. The test sand was a 1- to 2-darcy river-bank sand that closely approximated the Miocene producing formations in the U.S. gulf coast. In addition to providing data on strength and permeability of the consolidation process, the apparatus in Fig. 12.1 yields information about how deeply resin can penetrate the sand.

Another important consolidation testing apparatus (**Fig. 12.2**) is a large, 5-ft radial model. In this model, the sandpack is 1 ft thick, covers one-fifth of a circle (70° arc), and extends up to 5 ft from the wellbore. Experiments conducted with this model give information about the size and properties of the consolidated sand mass under radial-flow conditions.

The size and flow characteristics of the large radial model essentially duplicate the near-wellbore characteristics of a producing well. The evaluation program conducted in this model comes as

TABLE 12.1—COMMERCIAL PLASTIC CONSOLIDATION PROCESSES

Resin	Sand-Coating Method: Catalysis	Sand-Coating Method: Permeability	Resin Viscosity (cp at 70°F)	48-in. Linear Tests*: Initial Strength (psi)	48-in. Linear Tests*: Permeability Retention (%)	5-ft Radial Tests**: Initial Strength (psi)	5-ft Radial Tests**: Permeability Retention (%)	5-ft Radial Tests**: Consolidated Radius (in.)
Epoxy	Internal	Phase separation	5	3,500	70	2,500	75	41
	Internal	Overflush	15	6,000	55	3,500	70	43
	In situ	Overflush	15	6,000	60	6,000	75	32
Furan (including furan/phenolic)	In situ	Overflush (oil)	15	3,500	55	2,500	75	32
	In situ	Overflush (water)	15	4,000	30	2,500	40	27
Phenolic	Internal	Phase separation	10	2,000	65	2,000	60	40
	Internal	Overflush	25	2,000	60	1,500	50	37

*Acidized with HF/HCl.
**Acidized with HCl only.

close to reproducing actual field-treatment results as laboratory studies.

Field Data. Since the procedures were first developed, several thousand sand consolidation treatments have been performed and many studies conducted to determine their effectiveness. These studies identified not only the initial sand control success rate but also how this success changes with well production. The major findings of these studies are included in this chapter.

Summary of Evaluation

As described in Chap. 11, consolidation processes can be characterized by the type of resin and the method used to coat the sand. These factors provide the primary technical differences among the 15 to 20 plastic consolidation systems offered commercially. Each generic category may include several competing commercial systems, each with a unique set of properties depending on the exact resin composition, catalyst, and field placement procedure. For this evaluation, results considered typical of each generic category are reported; it is beyond the scope of this chapter to discuss each commercial system individually.

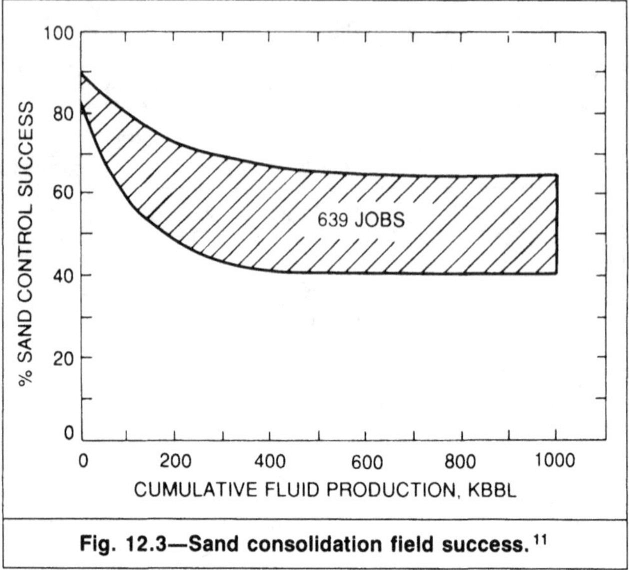

Fig. 12.3—Sand consolidation field success. [11]

Laboratory Data. Table 12.1 summarizes laboratory data for commercial consolidation systems by resin type and sand-coating method. Data on typical resin viscosity are given as a guide for comparing the ease with which the resins can be pumped. Laboratory consolidation data from two sources are presented: 48-in. linear tests on HF/HCl-acidized Brazos River sand and 5-ft radial tests on HCl-acidized Brazos River sand.[4]

In general, all commercial resins and coating methods in Table 12.1 can achieve sufficient initial strength and consolidated radius to control sand effectively.* For 48-in. linear sandpacks acidized with HF/HCl, epoxy resins produced the highest initial strength, followed by furans and then phenolics. In the radial tests, HF/HCl dissolved too much sand and created channels in the weakly stressed sandpack. Testing in the radial model was limited to sand acidized with HCl only. Radial results were similar to linear results, but as anticipated, the HF/HCl-acidized sand (linear tests) generally had higher strengths.

*Unpublished studies, Exxon Production Research Co., Houston (1971–78).

In terms of permeability retention, most processes retain sufficient flow capacity that well productivity will not be seriously impaired. Systems that include an aqueous overflush fluid retain permeability less efficiently owing to the unfavorable mobility ratio between the resin and the overflush, yet these systems have achieved acceptable results in field applications.

Field Results. Fig. 12.3 shows field results for more than 600 plastic consolidation treatments in the Gulf of Mexico. Sand-control success (as defined in Chap. 2) is plotted vs. cumulative fluid production per well (maximum of 1 million bbl). The data for several commercial systems plot as a band with an initial success rate of 80 to 90% and a long-term success rate between 40 and 65%. Overall, furan systems were near the top of the band, epoxies were near the middle, and phenolics were at the bottom.

However, before drawing any general conclusions from Fig. 12.3, note that no single resin system performed best in all individual fields. In some fields, phenolic resins performed better than either furans or epoxies. Results in a specific field are influenced by the formation grain size, mineralogy, and crude oil composition. A va-

TABLE 12.2—TREATMENT DESIGNS FOR EPOXY CONSOLIDATION SYSTEMS

	Internally Catalyzed Phase Separation		Internally Catalyzed Overflush		In-Situ Catalyzed Overflush	
Fluid	Ingredients	Approximate Volume	Ingredients	Approximate Volume	Ingredients	Approximate Volume
Acid	HCl HF/HCl Mutual solvent	50 gal/ft 100 gal/ft	HCl HF/HCl Mutual solvent	50 gal/ft 100 gal/ft	HCl HF/HCl Mutual solvent	50 gal/ft 100 gal/ft
Neutralizer	Water NH_4Cl NH_4HCO_3	150 gal/ft	Water NH_4Cl NH_4HCO_3	150 gal/ft	Water NH_4Cl NH_4HCO_3	150 gal/ft
Preflush	Diesel Alcohol Aromatic oil	300 gal/ft 300 gal/ft 100 gal/ft	Diesel Alcohol Aromatic oil	300 gal/ft 300 gal/ft 100 gal/ft	Mutual solvent Viscosifier Coupling agent	100 gal/ft
Resin	Epoxy resin Curing agent Diluent	150 gal/ft	Epoxy resin Curing agent Diluent	65 gal/ft	Epoxy resin Diluent	55 gal/ft
Spacer	—	—	—	—	Refined oil	30 gal/ft
Overflush	—	—	Refined oil	200 gal/ft	Refined oil Amine catalyst	250 gal/ft
Displacement	Diesel	Tubing	Diesel	Tubing	Diesel	100 gal/ft + tubing
Cure time	12 hours		12 hours		12 hours	

TABLE 12.3—TREATMENT DESIGNS FOR FURAN (OR PHENOLIC/FURAN) CONSOLIDATION SYSTEMS

	In-Situ Catalyzed Overflush (Oil)		In-Situ Catalyzed Overflush (Water)	
Fluid	Ingredients	Approximate Volume	Ingredients	Approximate Volume
Acid	HCl HF/HCl Mutual solvent	50 gal/ft 100 gal/ft	HCl HF/HCl Mutual solvent	50 gal/ft 100 gal/ft
Neutralizer	Water NH_4Cl NH_4HCO_3	150 gal/ft	Water NH_4Cl NH_4HCO_3	150 gal/ft
Preflush	Diesel + surfactant or mutual solvent	100 gal/ft	Brine + surfactant or mutual solvent	100 gal/ft
Resin	Furan (or Phenolic) resin Furfuryl alcohol Coupling agent	90 gal/ft	Furan (or Phenolic) resin Furfuryl reactive diluent Coupling agent	90 gal/ft
Spacer	Diesel + surfactant	40 gal/ft	Brine + surfactant	40 gal/ft
Overflush	Diesel + surfactant Organic acid catalyst	300 gal/ft	10% HCl + surfactant NaCl	300 gal/ft
Displacement	Diesel	Tubing	Brine	Tubing
Cure time	12 hours		12 hours	

riety of resin systems generally is needed; the challenge is to select the best one for each specific application. The following section details some field-treatment design aspects of the various commercial systems so that the engineer can judge which treatment alternatives address a particular consolidation problem.

Field-Treatment Design. The field-treatment designs for most commercial consolidation systems are similar, although the chemical composition used in each step can be unique. Basic treatment design information is given for the three resin types and sand-coating methods. The intent here is to identify the range of techniques used for the 15 to 20 commercial plastic consolidation systems. To apply a specific commercial system, a recommendation should be obtained from the supplier.

Typical ingredients and approximate treatment volumes for each type of plastic consolidation process are described. Treatment volumes are given in terms of the number of gallons to be injected into each foot of *perforated* interval (not sand thickness). A volume of 100 gal/ft will completely fill the pore space surrounding the perforated casing to a radius of about 3½ ft; a typical consolidated radius is 3 ft.

An acid treatment is generally used before a consolidation process to remove formation damage and to provide a clean sand surface for maximum resin adhesion. All the treatment-design tables show that an acid treatment precedes the consolidation chemicals. Acidizing consists of injecting HCl, followed by HF/HCl (with a mutual solvent), and then a neutralizer.

Epoxy Resin Systems. Commercial consolidation systems based on epoxy resin have been developed with all three sand-coating methods. **Table 12.2** summarizes the ingredients and approximate treatment volumes for each.

Regardless of the sand-coating method, the degree of bonding obtained with an epoxy resin is sensitive to the amount of water present in the pore space; hence, all epoxy systems include a water-removing preflush. For some systems, reservoir crude is displaced first with diesel and then water is removed with an alcohol. Alternatively, a mutual-solvent preflush can displace both water and

TABLE 12.4—TREATMENT DESIGNS FOR PHENOLIC CONSOLIDATION SYSTEMS

	Internally Catalyzed Phase Separation		Internally Catalyzed Overflush	
Fluid	Ingredients	Approximate Volume	Ingredients	Approximate Volume
Acid	HCl HF/HCl Mutual solvent	50 gal/ft 100 gal/ft	HCl HF/HCl Mutual solvent	50 gal/ft 100 gal/ft
Neutralizer	Water NH_4Cl NH_4HCO_3	150 gal/ft	Water NH_4Cl NH_4HCO_3	150 gal/ft
Preflush	Diesel or mutual solvent	70 gal/ft	Diesel + mutual solvent	80 gal/ft
Resin	Phenolic resin Coupling agent Base catalyst	120 gal/ft	Phenolic resin Coupling agent Base catalyst	80 gal/ft
Spacer	—	—	Diesel	80 gal/ft
Overflush	Diesel	80 gal/ft	Diesel + accelerator	80 gal/ft
Displacement	Diesel	Tubing	Diesel	Tubing
Cure time	24 hours		24 hours	

crude; the solvent may contain additives such as a coupling agent or a viscosifier.

Epoxy sand consolidation systems usually include either a bisphenol A or a bisphenol F epoxy resin; however, epoxy novolac resins have been used on a limited basis.[1,3-5,7] As manufactured, epoxy resins are very viscous; they require addition of a diluent, such as acetone or ethyl acetate, to be pumpable in field operations. To harden the resin, internally catalyzed systems contain a curing agent, typically a primary diamine. The hardening reaction begins as soon as the curing agent is added, but a typical design allows 2 or 3 hours of pumping time before the viscosity increases or the resin phase separates. The *in-situ* catalyzed resins do not contain a curing agent and will not harden until contacted by the catalyst.

The internally catalyzed phase-separation system is not overflushed; it is only displaced to the formation face and allowed to cure. The internally catalyzed and *in-situ* catalyzed overflush methods use a refined oil to establish permeability.[3-5] Either a motor-oil bright stock or a mineral-type oil will provide a favorable mobility ratio for opening pores to flow. The overflush for the in-situ catalyzed systems contains a few percent of a tertiary amine that causes the resin to harden.

Although 12 hours usually is required before workover operations can be resumed, the plastic may continue to cure and become stronger for 24 hours. High reservoir temperatures will promote a more rapid curing.

Furan (or Phenolic/Furan) Resin Systems. Furan resins are very reactive, and no practical method has been devised to catalyze them internally for sand consolidation. All commercial processes based on furan or furan/phenolic resins are catalyzed *in situ*. Oil and water overflushes are used. **Table 12.3** gives treatment designs for both techniques.

For the oil-overflushed systems, diesel plus a surfactant is used as the standard preflush. It reduces the residual water saturation and may promote wetting of the sand by the resin. Brine plus a surfactant is commonly used to preflush aqueous systems. Brine salinity is kept high to prevent water-soluble components in the resin from dissolving in the preflush. A mutual-solvent preflush can also be effective for either system.

Resin formulations can be based on either furan or phenolic technology. Regardless of the choice of resin, oil-overflushed systems commonly include furfuryl alcohol as a reactive diluent. For the aqueous systems, furfuryl alcohol is too water soluble, so another reactive diluent, such as furfuryl aldehyde, is used. The resin usually contains a coupling agent to promote good adhesion to the sand.

Spacers are used to separate the resin from the catalyst overflush. Diesel is used for the oil-overflushed systems, and high-salinity brine is common for the aqueous ones. A surfactant can be added to both.

The catalytic overflush for the oil systems consists of diesel plus an oil-soluble organic acid (or acid derivative). Because furan systems are very reactive, the catalyst concentration is typically small. Aqueous systems are generally catalyzed with HCl. Acid concentration ranges from 10 to 15%, and NaCl may be added to increase salinity and decrease the solubility of water-soluble resin components. In many cases, a surfactant is added to the overflush.

Furan resins cure very rapidly after exposure to the catalyst. A 12-hour (overnight) curing time is generally used, but in many cases workover operations could be resumed in less time.

Phenolic Resin Systems. In contrast to furan or furan/phenolic systems, plain phenolic resins polymerize slowly. Therefore, it is practical for commercial phenolic systems to be catalyzed internally. As **Table 12.4** shows, either phase separation or an overflush may be used to establish permeability.

Several options are available to prepare the sand for coating by the resin. A diesel preflush has been used successfully, and a mutual-solvent preflush can be used alone or in combination with diesel.

Resin formulations are all similar; they contain a phenolic resin, a coupling agent, and a base catalyst (e.g., sodium hydroxide). The polymerization reactions begin as soon as the catalyst is added to the resin. Working time before phase separation occurs or before the viscosity increases is typically 1 to 3 hours, which is adequate to pump the resin into the formation. The reactivity of a base-catalyzed phenolic resin is sensitive to the temperature in the reservoir sand. To control the reaction time, the resin ingredients are modified according to the reservoir temperature.

Surprisingly, the phase-separation system uses a small overflush volume to assist in developing permeability. The resin content of this system is quite high, and adequate compressive strength is achieved even though the pore spaces are not filled entirely with resin solution. The overflush systems include diesel to remove excess resin and establish permeability. A small amount of accelerator is added to the overflush to assist in polymerizing the resin.

Adequate compressive strength to resume workover operations is achieved within 24 hours.

Summary

The primary criteria for evaluating consolidation processes are compressive strength, permeability retention, aging characteristics, and field results. Based on the available laboratory and field data, Table 12.1 shows the compressive strength and permeability retention that can be anticipated from plastic consolidation systems, provided that they are placed in accordance with the treatment designs listed in Tables 12.2 through 12.4.

References

1. Richardson, E.A. and Hamby, T.W.: "Consolidation of Silty Sands With an Epoxy Resin Overflush Process," *JPT* (Sept. 1970) 1103-08.
2. Brooks, F.A.: "Consolidation of Dirty Sands by Phenol-Formaldehyde Plastic," *JPT* (Aug. 1971) 934-38.
3. Brooks, F.A. *et al.*: "Externally Catalyzed Epoxy for Sand Control," *JPT* (June 1974) 589-94.
4. Shaughnessy, C.M., Salathiel, W.M., and Penberthy, W.L. Jr.: "A New, Low-Viscosity, Epoxy Sand Consolidation Process," *JPT* (Dec. 1978) 1805-12; *Trans.*, AIME, **265.**
5. Treadway, B.R., Brandt, H., and Parker, P.H. Jr.: "Studies of a New Process To Consolidate Oil Sands With Plastics," *JPT* (Dec. 1966) 1537-43.
6. Murphey, J.R., Bila, V.J., and Totty, K.: "Sand Consolidation Systems Placed With Water," paper SPE 5031 presented at the 1974 SPE Annual Meeting, Houston, Oct. 6-9.
7. Bezemer, C. and Meijs, F.H.: "Development and Field Application of the Process for Sand Consolidation Employing EPOSAND Resin," paper SPE 1591 presented at the 1966 SPE Annual Meeting, Dallas, Oct. 2-5.
8. Hamby, T.W. and Richardson, E.A.: "New Sand Control Process Works Well in Dirty Sands," *World Oil* (June 1968).
9. Muecke, T.W.: "Factors Influencing the Deterioration of Plastic Sand Consolidation Treatments," *JPT* (Feb. 1974) 157-66.
10. Rensvold, R.F.: "Sand Consolidation Resins—Their Stability in Hot Brine," *SPEJ* (April 1983) 238-48.
11. Powell, J.L.: "Analysis of Sand Control Experience for Offshore Wells in the Gulf of Mexico," Exxon Co. U.S.A. (Sept. 1976).

SI Metric Conversion Factors

bbl	× 1.589 873	E−01	=	m^3
ft	× 3.048*	E−01	=	m
°F	(°F−32)/1.8		=	°C
gal	× 3.785 412	E−03	=	m^3
in.	× 2.54*	E+00	=	cm
md	× 9.869 233	E−04	=	μm^2
psi	× 6.894 757	E+00	=	kPa

*Conversion factor is exact.

Chapter 13
Field Procedures for Plastic Consolidation

Introduction

Field procedures for plastic consolidation can significantly affect job success. One goal is to ensure that all resin is flushed, or displaced, from the wellbore tubulars and equipment into the formation. Serious production problems can arise if even a portion of the resin hardens inside the wellbore.

The other important goal of the field procedure is to achieve uniform resin distribution in the formation around the well. Resin distribution is heavily influenced by the number of perforations open for flow. Maximizing the number of open perforations requires close attention to perforating and perforation-washing techniques. Prepacking the formation is frequently required to fill any voids or channels that may distort the resin distribution.

Tubular Integrity

Sound wellbore equipment is absolutely essential for successful plastic consolidation treatments. Any conditions that allow consolidation chemicals to enter areas of the wellbore other than the interval to be treated not only will result in an unsuccessful job, but also can cause the hole to be junked.

An unrepaired leak in the wellbore tubular goods is the surest way to "glue" a workstring in the well. Any damage allowing fluids to leak from the workstring/tubing annulus will permit fluids to exit the wellbore at some point other than the perforations. Resin (and subsequently the catalyst) will flow *up* the annulus, causing the bottom portion of the workstring to be glued in the well. The current record for workstring glued this way is 3,000 ft.

Many conditions can cause leaks. Obviously, such problems as split casing, split tubing, or a failed squeeze-cement treatment of an old perforated interval are major sources. If the well is on gas lift, then the valves are usually replaced with dummies to prevent damage. Therefore, each gas-lift mandrel is a potential leak site. At the very least, the gas-lift annulus should be killed and a gauge used to monitor its pressure during the treatment. Preferably, the tubing string should be pulled and the gas-lift mandrels removed.

The surface-controlled subsurface safety valve is another potential leak source. This, too, must be pulled and replaced by a dummy to prevent resin from entering the control lines. The pressure of any annular space associated with the safety valve should be monitored to detect leaks.

If a well has sanded up, then it is frequently convenient to test the integrity of the wellbore tubulars at that time. Because no fluid can be injected through sanded-up perforations, pressure testing of the casing or tubing can easily be performed.

Perforating

The relationship between perforating practices and sand-control success has been the topic of many studies. The results of these studies can be used to choose the best perforating technique for performing a plastic consolidation.

Shot Density. Every study of plastic consolidation has demonstrated conclusively that increasing the shot density also increases the chances for sand-control success. **Fig. 13.1** shows the results of such a study for the Gulf of Mexico. Long-term consolidation success is clearly improved by 20% when a shot density of 4 shots/ft instead of 1 or 2 shots/ft* is used.

Why might a high shot density improve consolidation success? The answer is the distribution of resin around the well. If half the perforations are accepting the injected fluid and shot density is only 1 shot/ft, several feet of interval may receive no consolidating fluid. In theory, these plugged perforations would be covered by resin coming from adjacent perforations. However, the farther apart these adjacent perforations are, the less likelihood of consolidating around all of them. At 2 shots/ft, and especially at 4 shots/ft, the vertical distance between perforations actually accepting the consolidating resin becomes small (1 ft or less).

Shot densities exceeding 4 shots/ft generally are not used with plastic consolidation systems. The perforations are left open so that they do not restrict fluid production. This situation is in contrast to gravel packing, where the perforations are filled with sand. Shot densities for gravel packing can be 8 to 12 shots/ft to reduce flow resistance.

Hole Size. In plastic consolidation, perforation size is not nearly as important as the number of holes. Large holes are desirable, however, to increase the flow area and to reduce the local drawdown stress. In gravel packing, large perforations are required because they always fill with gravel. In consolidation, the perforation tunnels are left open, so their diameters are of less importance.

More important than the hole size created by the perforating gun is the perforation penetration. Firing 4 shots/ft does not always result

*Unpublished report, Exxon Production Research Co., Houston (1965).

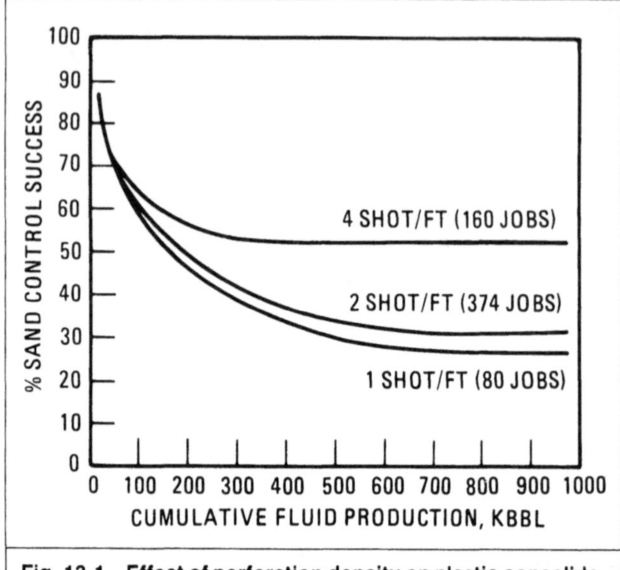

Fig. 13.1—Effect of perforation density on plastic consolidation success.

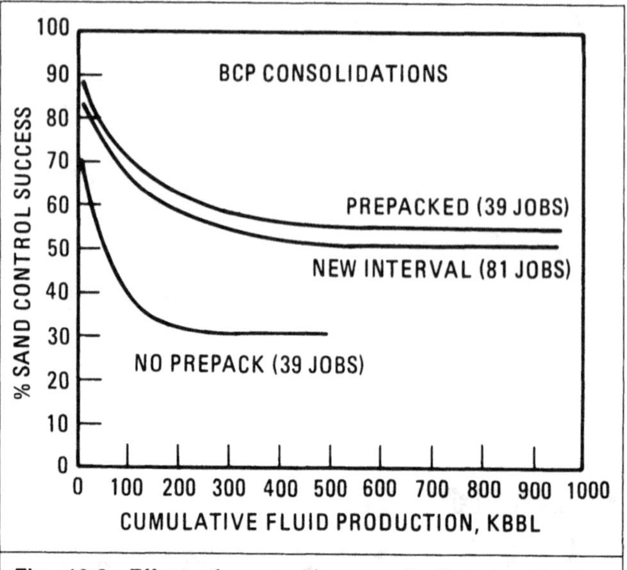

Fig. 13.2—Effect of prepacking on plastic consolidation success.

in four shots that actually penetrate the casing and communicate with the formation. The perforating tool chosen should achieve the most predictable penetration success, even if a certain amount of diameter is sacrificed.

Prepacking

Prepacking a formation before consolidation fills the void spaces behind the casing caused by sand production. Void spaces in the near-wellbore region can lead to resin channeling and thus uneven plastic distribution. If void spaces become large enough, the resin and catalyst will segregate. The less dense catalyst (oil) will float over the resin, leaving a significant portion of the resin uncured.

The previous discussions of gravel-pack productivity seem to indicate that prepacking before sand consolidation would adversely affect well productivity because of consolidated prepack material in the perforation tunnels. However, this does not seem to be the case because at the time of the consolidation treatment, the prepack material is primarily in voids outside the casing. The consolidated prepack material remaining in the perforation tunnels has not presented a problem. Reasons for these observations include the following:

1. The prepack material is washed out of the wellbore before the consolidation treatment and much of that in the perforations is removed.
2. The sand consolidation fluids are viscous and tend to displace remaining prepack material through the perforation during the treatment.
3. After the resin is pumped through the perforations, many PV's of overflush and catalyst fluid are also pumped through the perforations. Prepack material remaining in the perforations would be weakly consolidated because of low resin saturations that result from the large volume of overflush- and catalyst-fluid displacement.
4. Subsequent fluid production flowing at high velocities through the perforations tends to remove the weakly consolidated prepack material from the perforation tunnels.

Correlation With Success. The correlation between prepacking and consolidation success has been firmly established by several field-lifetime studies. **Fig. 13.2** shows the results of one Gulf of Mexico study. To eliminate as many variables as possible, the scope of this particular study was limited to intervals perforated with 2 shots/ft and consolidated with a phenolic resin. All were "old" intervals that had produced sand before treatment. The results clearly show that prepacking old intervals before consolidation has significant benefits.

As Fig. 13.2 shows, sand-control success for old intervals after prepacking is equivalent to success in "new" intervals that have not produced sand. The prepacking operation apparently can restore the sand distribution around a sand-producing well to its undisturbed state.

Procedure. Prepacking a formation is uncomplicated and easy. Sand is suspended at a low concentration in a low-viscosity carrier fluid and pumped into the formation until a sandout occurs. Ordinarily, the well is reverse circulated clean and the procedure repeated until a second sandout occurs.

A 40/60-U.S.-mesh sand is used for most sandpacking operations. This size of sand is easily suspended in the carrier fluid and readily consolidated by resin. Large sand grains do not have enough contact points to be strongly consolidated. The 40/60-U.S.-mesh sand is typically 10 times more permeable than the formation sand; therefore, no production impairment will result from prepacking.*

Diesel or brine is the usual choice for a carrier fluid. Viscosifying additives are not needed because 40/60-U.S.-mesh sand is readily transported by either fluid. Also, viscosity must be kept low so that the fluid will easily "leak off" into the formation, leaving the sand packed tightly in void spaces. Low viscosity is important also because the carrier fluid must be displaced by the preflush and resin during consolidation.

To prevent a premature sandout at the perforations, the sand concentration is kept low, normally 0.25 to 0.50 lbm/gal. Typical pumping rates are ½ bbl/min or higher. Sanding out is considered to have occurred when the injection pressure increases by 200 to 300 psi during treatment. This pressure can be monitored on the injection line or, preferably, on the workstring/tubing annulus. High packing pressures should be avoided because they can lead to grain crushing and formation fracturing.

A downstream sand injector is frequently used to facilitate sandpacking. This device allows sand and carrier fluid to be mixed downstream of the fluid pump. The sand injector reduces the amount of sand crushing and prevents pump abrasion by the sand.

Injectivity Testing

The ultimate goal of any injectivity testing procedure is to ensure that all the consolidation chemicals can be pumped without exceeding the formation's fracture pressure. The injectivity response must be determined in advance because, once pumped, consolidation chemicals cannot be recovered and reused.

Exceeding the formation fracture pressure is not permissible on a matrix-type treatment. In such treatments (e.g., consolidation), the goal is for fluid to reach every pore space in the near-wellbore region. Exceeding the fracture pressure creates a channel that leads the treating fluids away from the area immediately surrounding the

*Unpublished procedures, Exxon Production Research Co., Houston (1970).

perforations. The consolidation treatment probably will not succeed because much of the consolidation chemicals are at substantial distances from the perforations where sand control is needed.

Testing Procedure. A single pressure reading cannot be used as a true measure of well injectivity. The difference in pressure under static and injection conditions should be used because it is an accurate assessment of the amount of pressure increase associated with fluid injection.

To measure injectivity, the pressure recorded under static conditions is subtracted from the pressure recorded during pumping. Diesel is usually selected as the injection fluid for testing oilwell injectivity; brine is used for gas and water injection wells. A 0.5-bbl/min pumping rate is typical, but it can be adjusted according to interval length and formation permeability.

The location of the pressure measurement is also important. The pressure gauge should be located on a static annulus kept full of liquid. The static annulus between the workstring and the tubing is usually chosen. Pressure measurements on the injection line reflect the frictional pressure drop of the tubing and the formation injectivity. The frictional pressure drop can be 1,000 psi or greater in a 1-in. workstring. By contrast, the static-annulus fluid column gives an accurate measure of the pressure response at the formation face.*

Workstring Treatment Pressure. Some consolidation fluids are four to five times more viscous than diesel. A pressure increase of 300 to 500 psi recorded during injectivity testing with diesel will therefore lead to a 1,200- to 2,500-psi increase at the formation face when consolidation fluids are pumped *at the same rate.* Under most circumstances, this injectivity would be sufficient to allow pumping of the consolidation chemicals. However, the static pressure plus the injection pressure during consolidation must not exceed the fracture pressure:

$$p_{swh} + (1{,}200 \text{ to } 2{,}500 \text{ psi}) < p_{fwh},$$

where

p_{swh} = static wellhead pressure, psi, and
p_{fwh} = fracture pressure at the wellhead, psi.

If the calculation shows that the fracture pressure would be exceeded at any time, the first option is to reduce the pumping rate for the consolidating chemicals and recalculate the pressure during injection. If the injection pressure is still too high, some type of perforation cleaning, such as acidizing, should be attempted. Knowledge of the native reservoir permeability is important in such cases because the entire problem may be low reservoir permeability.*

Bullhead Treatment Pressure. A sand consolidation treatment is sometimes done without a workover rig. In a bullhead treatment, the fluids are pumped directly down the production tubing. When this technique is used, there is no static-annulus pressure to reflect conditions at the formation face. The use of pressure differences is not reliable in bullheading because the difference between static and pumping wellhead pressures always includes a tubing pressure drop of a few hundred psi.

One common criterion for evaluating well injectivity in bullheading is to compute the pressure gradient during pumping at 0.5 bbl/min. If this gradient is less than half the fracture gradient, then the well has sufficient injectivity to proceed with the consolidation:

$$p_{wh}/D_{tV} + g_{fl} < \tfrac{1}{2} g_f,$$

where

p_{wh} = wellhead pressure, psi,
D_{tV} = true vertical depth, ft,
g_{fl} = fluid gradient, psi/ft, and
g_f = fracture gradient, psi/ft.

Although this equation has found widespread application, its use requires some knowledge of the reservoir pressure. In highly pressured reservoirs, injecting at half the fracture gradient may be impossible even though nothing is wrong with the well injectivity.

Acidizing

Acidizing is a method for removing clay, scale, and other particulate materials from the formation sand. Plugging of perforations by particles suspended in the workover fluid or loosened from the walls of the casing or tubing is common in workover operations. The particles impair injectivity (and subsequent productivity) and must be removed before consolidation. The presence of these damaging solids can usually be identified during an injectivity test.

In addition to improving injectivity, acidizing benefits a consolidation treatment by cleaning the sand-grain surfaces. Fresh silica is exposed when clay, feldspar, and other minerals are removed from the pore space. Maximum bond strength is developed between the consolidation resin and a freshly etched silica surface. Thus, even when no injectivity problem exists, an acid treatment is frequently performed to improve the strength of the consolidated sand.

Acid Treatments. To dissolve clays and other particular material, a mixture of 3% hydrofluoric (HF) and 12% hydrochloric (HCl) acids is used. This acid mixture, however, is rarely used by itself because many of the products produced by reaction with clay form insoluble precipitates in the presence of sodium ions. To displace the native brine, an HCl acid preflush always precedes the HF/HCl mixture into the formation.

Acid Volume. Typical volumes for an acid treatment are 50 gal/ft for 15% HCl and 100 gal/ft for 3% HF/12% HCl. The exact acid volume should be chosen on the basis of field experience in the area.

Acid can be injected through a workstring or bullheaded down the production tubing. If a workstring is used, the end should be positioned at the perforations and reciprocated across the perforated interval during acid injection. This technique will not ensure that the acid is injected uniformily over the entire interval. However, if the acid is flowing into a thief zone at the top of the perforated interval and if the end of the tubing string is above the interval, the lower perforations may not be treated effectively. Reciprocating the end of the workstring across the completion interval would at least subject the lower interval to the live acid. The same would be true if the lower interval were accepting the majority of the acid and the end of the workstring were located below it.

Acid Additives. HCl and HF corrosiveness may be controlled by an appropriate corrosion inhibitor. The effectiveness of such inhibitors depends on temperature and exposure time. Inhibitor concentrations of a few tenths of 1% are common for acids used in conjunction with sand consolidation.

Emulsion formation can occur when a low-pH fluid (acid) is mixed with an oil (crude or diesel). These emulsions are readily stabilized by the fines released or produced during the acidizing process. To combat this problem, a mutual solvent should be added to the acid. Other surfactants must be used cautiously because of possible adverse effects on sand wettability.

When a highly asphaltic crude is encountered, a spearhead of an aromatic solvent can be used effectively ahead of the acid treatment. Typical solvents include toluene and xylene. A spearhead of a dispersion of the aromatic solvent in 15% HCl also has been used.*

Neutralizer. In ordinary acidizing situations, the HF/HCl mixture is usually followed by an afterflush of either HCl or diesel. When acidizing is performed with a plastic consolidation treatment, however, the acid should be followed by a neutralizing solution to remove residual acidity from the sand. Residual acidity can be responsible for poor consolidation in two ways. First, when an acid-catalyzed consolidation plastic is used, residual acidity may cause premature hardening of the resin with a subsequent loss in permeability. Second, when a base-catalyzed plastic is used, residual acid can react with the catalyst and prevent hardening of the resin.

Removal of Acidity. A neutralizer can remove residual acidity through two mechanisms. The most important is the miscible displacement of the acid from the pore space. This mechanism may

*Unpublished procedures, Exxon Production Research Co., Houston (1970).

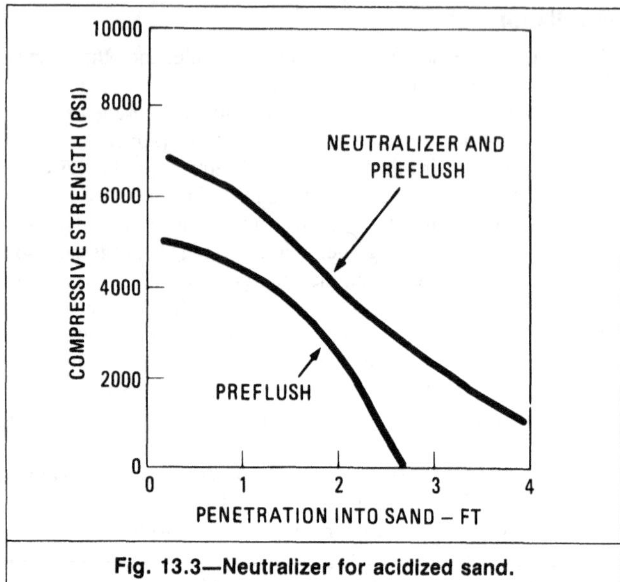

Fig. 13.3—Neutralizer for acidized sand.

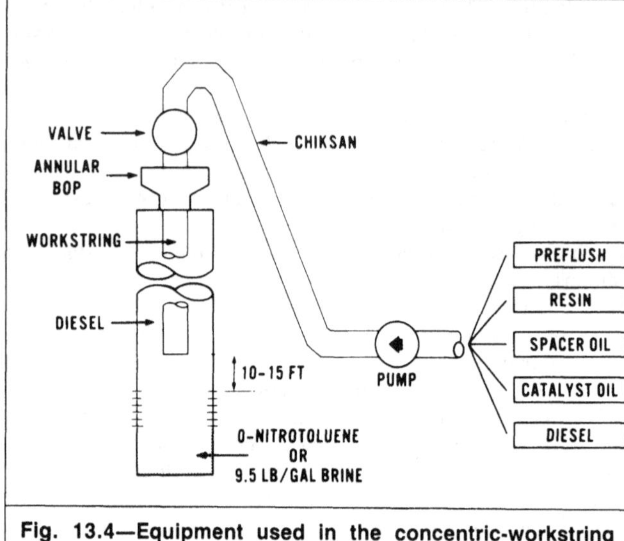

Fig. 13.4—Equipment used in the concentric-workstring technique.

be thought of as a dilution process. The second mechanism is for the neutralizer to react chemically with residual acid.

Fig. 13.3 illustrates the improved performance of a consolidation treatment after neutralization. An epoxy resin with a base catalyst was used in this test. The residual acid limited penetration of the consolidation to 2½ ft by deactivating a portion of the catalyst. When a neutralizer was used ahead of the preflush, the catalyst could penetrate 4 ft into the reservoir sand. Better catalysis was achieved over the entire length of consolidation, as evidenced by the higher compressive strength when the residual acid was neutralized.

Ingredients. Neutralizer capacity can be provided by either ammonium bicarbonate (NH_4HCO_3) or ammonium hydroxide (NH_4OH). These bases can be used interchangeably to provide a neutralization capacity of 0.25 mol/L. Because of the difference in molecular weight, the neutralization capacity can be achieved with either 2wt% NH_4HCO_3 or 1wt% NH_4OH.

The water used to dissolve these bases must not contain appreciable amounts of calcium or magnesium ions; otherwise, an insoluble precipitate will form during mixing. This precipitate has at times been great enough to plug the perforations and prevent further chemical injection. When the use of hard water is unavoidable, the neutralizer solution should be either filtered or mixed well in advance so that the precipitate may settle out.

In addition to water and base, which are essential to the dilution and neutralization mechanisms, a neutralizer must also possess sufficient ionic strength to prevent swelling of undissolved clays. This requirement is met by including 3 to 4% ammonium chloride (NH_4Cl) in the solution. Sodium or potassium chlorides are not useful because they form insoluble precipitates with spent mud acid.

The volume of neutralizer must be sufficiently large to displace residual acid from the pore space. The PV around the wellbore to be neutralized is approximately 200 gal/ft of perforated interval. To be effective, a minimum volume of 100 gal/ft of neutralizer is needed. Ordinarily, however, a volume of 150 gal/ft is used to ensure effectiveness. The cost of the neutralizer typically is low and therefore is not a major treatment expense.*

Consolidation Fluid Placement

The procedure used to place (or inject) consolidation fluids is important to job success. Consolidation involves a chemical reaction among several injected fluids. For the results of this reaction to be predictable, the reactants must arrive at the formation face uncontaminated. Because of the fluids' high costs, it is not feasible simply to increase consolidating-fluid volume to account for loss or contamination. Several unique injection procedures have been developed to deal with the placement problem. Although each is different, all were designed with three common objectives: to control fluid flow in the wellbore, to segregate reactive components, and to protect wellbore tubular goods against excessive plastic buildup.

Fluid Control. When resins that harden into insoluble plastics are used, the placement technique must guarantee that all the consolidating fluids are injected only into the formation. There must be no possibility that these fluids can accumulate in the rathole or in the workstring/casing annulus. Provided that there are no leaks, wellbore fluid flow can be controlled by correct selection of the densities of the fluids in the annulus and rathole.

Fluid Segregation. Good fluid segregation begins with having the consolidating chemicals arrive at the wellsite uncontaminated. Inside the well tubing, fluid segregation is maintained with either a fluid spacer or a mechanical separation plug.

Fluid spacers are widely used for chemical segregation. They are convenient because they require no additional mechanical equipment and cannot become stuck in the flowlines or tubing. The actual amount of separation achieved by each barrel of spacer depends on the tubing diameter. While 1 bbl of spacer fluid may be adequate in a 1-in. workstring (1,100 ft/bbl), it would certainly be inadequate for a bullhead treatment in 3½-in. tubing (120 ft/bbl). Typically, bullhead treatments are designed with 5 to 20 bbl of spacer fluid.

Brine and diesel are the two most common spacer fluids. A refined oil used to carry the catalyst in an *in-situ* catalyzed process is also used (without a catalyst) as the spacer between the resin and catalyst. Regardless of the fluid chosen, no spacer acts in a completely inert manner. Excessive spacer volumes tend to dissolve or wash a portion of the resin from the surface of the sand grains. Therefore, it is advisable to use the minimum volume of spacer consistent with achieving adequate fluid segregation.

Protection of Wellbore Tubing. The buildup of a thick layer of resin inside the well tubing can cause serious problems during future workovers. Low-viscosity (10- to 20-cp) resins that drain readily from the tubing walls minimize resin buildup. With higher-viscosity (50- to 100-cp) resins, a mechanical wiping device must be used to remove resin from the tubing walls. Pumping the consolidation treatment through a concentric workstring is, of course, the most successful method for preventing resin buildup inside the production tubing.

Concentric Workstring

Perhaps the most versatile means of placing consolidation fluids is with a concentric workstring. One advantage is the small fluid volume needed to fill the workstring. This allows precise injection

*Unpublished procedures, Exxon Production Research Co., Houston (1970).

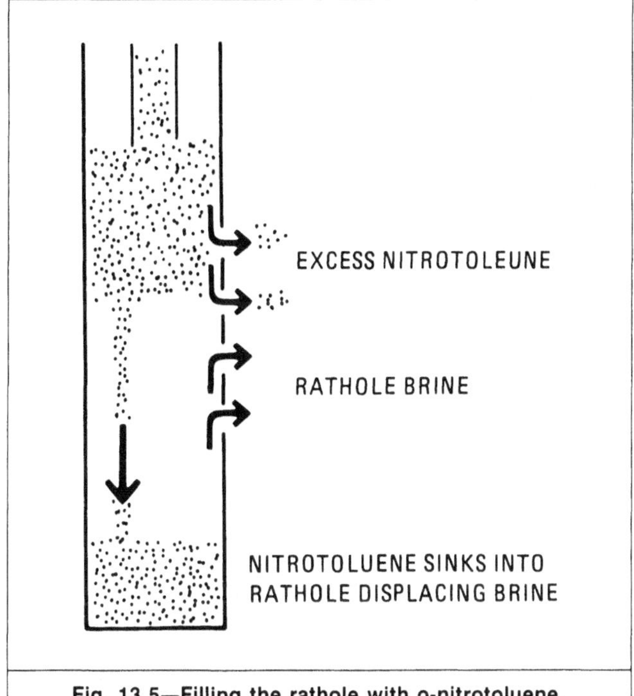

Fig. 13.5—Filling the rathole with o-nitrotoluene.

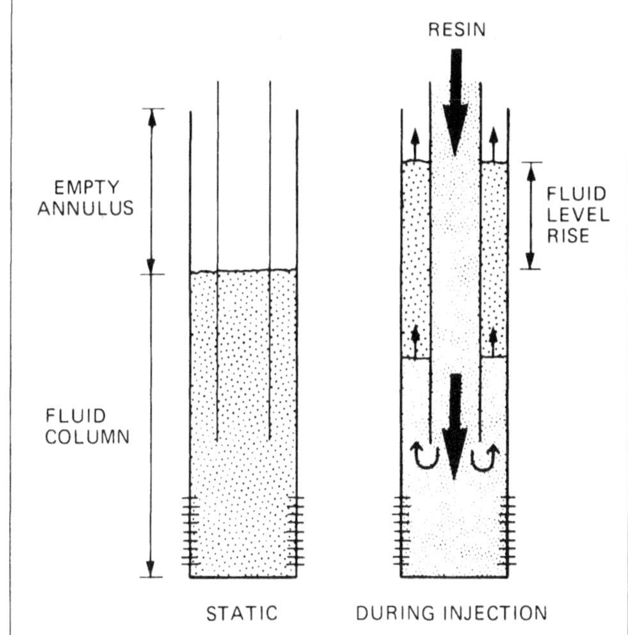

Fig. 13.6—Resin flowing around workstring when annulus will not stand full.

of the small chemical volumes usually encountered in consolidation treatment. **Fig. 13.4** shows the equipment layout for a typical concentric-workstring consolidation treatment.

Equipment. The primary piece of wellbore equipment used in a consolidation procedure is the workstring. A typical concentric workstring has a 1.315-in. OD and weighs 2.25 lbm/ft. Integral-joint tubing with internally flush joints is preferred. The design of these tubing joints allows a fluid spacer to flush resin from the workstring efficiently. A 1 ¼-in. workstring with standard couplings may also be used if these special joints are unavailable. A test sub is usually run on the bottom of the workstring to allow pressure testing after the string is in the well. The surface equipment through which the consolidating fluids are pumped should be kept as simple as possible. Pump suction should come directly from the chemical tanks and not through the rig suction manifold. Similarly, the pump discharge should go directly to the top of the workstring without passing through a complex manifold.

Because many of the consolidation fluids contain organic solvents, they can weaken rubber products. For this reason, only steel lines (not rotary hose) should be used on the discharge side of the pump. Because some workstring movement is required during a treatment, the workstring should be connected to the injection line from the pump by flexible joints. A maximum of 30 ft of workstring movement is generally all that is required.

To facilitate workstring movement during the job, a tubing stripper [or annular blowout preventer (BOP)] should be used on the wellhead. This equipment is supplemental to the customary ram-type BOP (not shown in Fig. 13.4). The joints in the workstring should be spaced by pup joints so that its end may be raised from 5 to 10 ft above the perforations *without* stripping a coupling through the rubber.*

Rathole Fluid. Because a plastic consolidation resin will cure to form a hard plastic, adequate control must be maintained over fluid flow in the wellbore. Two areas of immediate concern are the rathole and the annular space between the workstring and casing.

The density of a typical consolidation resin is 9.0 to 9.5 lbm/gal. Because it is more dense than usual wellbore fluids, resin tends to sink into the rathole below the perforations, flushing the original contents of the rathole (usually brine) into the formation and resulting in two detrimental consequences. First, the rathole is left full of resin, which can become partially catalyzed during consolidation. The partially cured resin can complicate future workover operations if it remains in the well. Second, and perhaps more seriously, brine displaced from the rathole by resin may contaminate the formation sand. The remedy is to fill the rathole with a fluid that is denser than the resin and thus prevent displacement.

Two techniques are available for filling the rathole with a dense liquid. First, a dense brine (9.5 to 9.8 lbm/gal) can be circulated into the rathole with a workstring. Dense brine has been used successfully many times in the field, but the process is somewhat time-consuming and is easily omitted. Dense brine is satisfactory for many consolidation systems but is not sufficiently reliable for use with epoxy resins that are water-sensitive.

In an epoxy consolidation process, the introduction of water from the rathole during resin injection reduces the strength of the final consolidation and can possibly cause the entire treatment to fail. In this case, a more reliable technique is essential: filling the rathole with o-nitrotoluene. In this technique, o-nitrotoluene, a dense (9.66-lbm/gal) organic liquid, is pumped to the perforated interval. As **Fig. 13.5** illustrates, a portion of the o-nitrotoluene sinks into the rathole because of gravitational forces, and the excess is injected through the perforations into the formation. The brine flushed from the rathole is similarly injected into the formation.[1]

To minimize the amount of o-nitrotoluene required, the rathole length should be limited to 25 ft. A volume of 300 gal pumped continuously at ¼ to ⅜ bbl/min will fill that length. Thus, it is unnecessary to stop pumping during the rathole-filling time. During pumping, the rathole fills with o-nitrotoluene and the excess material is injected into the formation sand.

Current safety information indicates that o-nitrotoluene should not be used with consolidation systems that include organic acid chlorides as catalysts.

Annulus Fluid. Consolidation fluids must be kept from entering the annulus between the workstring and casing in the region above the perforations. If the annulus will not stand full of fluid, then injection down the workstring into the formation will cause the fluid level in the annulus to rise. As **Fig. 13.6** shows, this condition can allow some of the resin to flow *up* the annulus after leaving the workstring. If pumping is now stopped—for instance, to clean the resin from the pump—the fluid level in the annulus will sink to its static level. In the meantime, the workstring's exterior has become coated with resin. When pumping is resumed for overflush injection, catalyst can flow up into the annulus, hardening the resin on the workstring, which can glue several joints of pipe into the well.

*Unpublished procedures, Exxon Production Research Co., Houston (1970).

To prevent this occurrence, the annulus can be filled with a fluid (usually diesel) with density low enough to allow it to stand full. When the bottomhole pressure (BHP) is so low that the annulus will not stand full of diesel, a retrievable packer should be run on the workstring. Customarily, about 30 ft of tailpipe is used below the packer. Having a full annulus or running a packer eliminates fluid-level rise in the annulus.

Even with the annulus standing full (or a packer run), there is a possibility of fluid movement if one of the consolidating fluids is less dense than the liquid in the annulus. Completely filling the annulus with a liquid less dense than the least dense consolidating fluid eliminates movement of annular fluid above the workstring. For consolidation systems that include an organic solvent preflush and/or overflushing fluid, diesel should be used to fill the annulus. For systems designed with aqueous preflush and overflush chemicals, a low-density brine may be used in the annulus if BHP permits.*

Procedure. After the well is prepared, the actual consolidation treatment begins. If the rathole is to be filled with dense brine, it should be done now. Next, the workstring is pulled to the top of the perforations and the well reverse circulated full of diesel. An injectivity test should be run to determine whether an acid treatment will be required. If needed, the well should be acidized and neutralized. Finally, the workstring should be pressure tested before the consolidation chemicals are pumped.

The workstring should be positioned 10 to 15 ft above the perforated interval. Most consolidation procedures are straightforward, requiring only that the chemicals be pumped in the correct sequence. Tanks with the proper amount of each chemical ingredient should be on location and clearly identified. The pump suction hose is moved sequentially to each tank.

The small workstring volume makes the concentric method ideal for placing small fluid volumes. Nonetheless, there is a limit on how small the volume can be. Given the waste and loss that occurs during the mixing and transportation of the consolidation chemicals, plus the amount left in tanks and suction hoses, the minimum practical volume is that necessary to consolidate 4 ft of interval. Shorter intervals can be treated, but the chemical volume for a 4-ft zone should be used.

Because of the small workstring volume, fluids injected at ambient conditions reach the bottom of the well quickly. The pumping rate must be slow enough to allow the consolidating fluids to warm up on the way to the bottom. A minimum of 100°F bottomhole injection temperature is desirable for most consolidation systems. A pumping rate of 0.3 to 0.5 bbl/min will provide sufficient residence time in the workstring for the 100°F minimum to be reached. For larger-diameter tubing, as in bullhead injection, a higher injection rate may be used, but a minimum residence time of 20 to 30 minutes should be maintained. Waiting for the formation to warm up after injection is another alternative; however, changes in the ambient and reservoir temperatures complicate this somewhat. Adhering to the above pumping schedules has given satisfactory results.

Because most of the acid-catalyzed systems polymerize very rapidly when contacted by a catalyst, it is necessary to flush out the injection pump thoroughly with spacer fluid between resin and catalyst injection. Sometimes the flushing involves breaking lines loose and draining excess resin from the pump. With the slower-curing base-catalyzed resins, the sweeping action of flowing the spacer fluid though the pump cleans the system sufficiently, so the problem of resin hardening in the pump does not arise.

Workstring movement during treatment is very limited. Once all the resin has exited the bottom of the workstring and spacer fluid begins to enter the perforations, the workstring is usually picked up 5 ft. This ensures that its end is not in contact with the resin that coats the casing wall between the workstring and the perforated interval. At no time after pumping the consolidation resin should the workstring be lowered.

After all the consolidating fluids have been injected, the workstring is picked up an additional 20 to 30 ft. The well is then shut in with this configuration for the required resin curing period. At the end of this period, an injectivity test is run, the well circulated with kill-weight fluid, and the workstring removed from the well.*

Uses. Because it is the most effective method for complete segregation of consolidation fluids, the concentric-workstring method is probably the best injection technique. It should be used whenever a rig is on location. There appears to be no advantage in skidding the rig and performing a bullhead treatment because only 1 day of rig time is saved.

Conventional Workstring

For the purposes of this discussion, a conventional workstring is considered to be 2⅜- or 2⅞-in. drillpipe or tubing not held by a packer at the bottom of the well. Because the techniques are so similar, most of the statements made for the concentric workstring also apply to the conventional workstring.

Equipment. The downhole wellbore configurations of conventional and concentric techniques are identical (see Fig. 13.4). The rathole must be filled with a dense brine or o-nitrotoluene, and the annulus must stand *full* of diesel (or a packer must be run). The end of the tubing should be placed 10 to 15 ft above the perforations at the start of the job.

Surface equipment for a conventional-workstring procedure is similar to that shown in Fig. 13.4. Because pipe movement is the same for both types of workstrings, similar tubing stripping equipment (or annular BOP) and flexible joints are needed.

Procedure. The placement procedure for conventional and concentric workstrings is essentially identical. However, the large internal volume of a conventional workstring—about 30 bbl or 1,200 gal—necessitates a slight change in job approach. The preflush, the resin, the spacer, and a portion of the overflush fluid are pumped into the workstring before the preflush ever reaches the formation sand. Because all the chemicals are in the tubing simultaneously, chances are greater that adjacent chemicals will mix, especially the resin and preflush. The dense resin is placed above the lower-density preflush in the workstring. This condition is unstable; the resin tends to fall through the preflush as the fluids are being pumped to bottom.

This mixing depends primarily on time spent in the tubing and not on flow rate. For this reason, a pumping rate of 0.5 to 1.0 bbl/min is recommended to minimize the time required to pump the chemicals to bottom. Because of fluid mixing, small chemical volumes cannot be successfully pumped through a conventional workstring. For most applications, the volume required to consolidate a 6-ft interval is the minimum amount of fluid to pump. A shorter interval can be treated, but the chemical volume for a 6-ft interval should be used.

Another result of a large workstring volume is that all the chemicals are essentially committed (pumped into the tubing) before the first chemical reaches the formation. Should any injectivity problems arise during the treatment, all the chemicals would have to be circulated out of the well and disposed of at the surface. For this reason, it is especially important to obtain good fluid injectivity before the treatment begins.*

Uses. Because of its large internal volume, a conventional workstring is not as suitable as a concentric workstring for plastic consolidation. In addition, a conventional rig is more costly than a concentric one, especially offshore. For some onshore applications, however, it is more economical to use the well tubing with a conventional rig instead of renting a concentric workstring.

Bullhead

Bullheading refers to pumping the consolidation chemicals directly down the permanent production tubing or casing. The well can be either a tubingless or a conventional completion with a pipe diameter of 2⅜, 2⅞, or 3½ in. The chemicals are pumped through the Christmas tree because there is no rig.

*Unpublished procedures, Exxon Production Research Co., Houston (1970).

Equipment and Procedure. Fig. 13.7 shows the equipment requirements for the bullhead placement method. Because it is impossible to circulate the rathole full of brine without a rig, o-nitrotoluene injection can be used to fill the rathole. To minimize the amount of o-nitrotoluene needed, the rathole depth should be limited to 25 ft. Current safety information indicates that o-nitrotoluene should not be used with consolidation systems that include organic acid chlorides as catalysts.

In conventional completions, the annulus between the tailpipe below the packer and the casing should be filled with diesel. Because there is no rig to circulate this fluid into place, it can be positioned only by pumping the diesel to the bottom and letting gravity direct it to the proper location. Pumps and manifolds are the only surface equipment required for a bullhead treatment. The fluids are essentially placed by pumping the consolidation chemicals in the correct sequence. As in all placements, the pumping rate must be controlled to prevent fracturing of the formation. The same tubing-volume considerations applicable to conventional-workstring placement apply to bullhead treatments. For this reason, a pumping rate of 0.5 to 1.0 bbl/min is commonly used.*

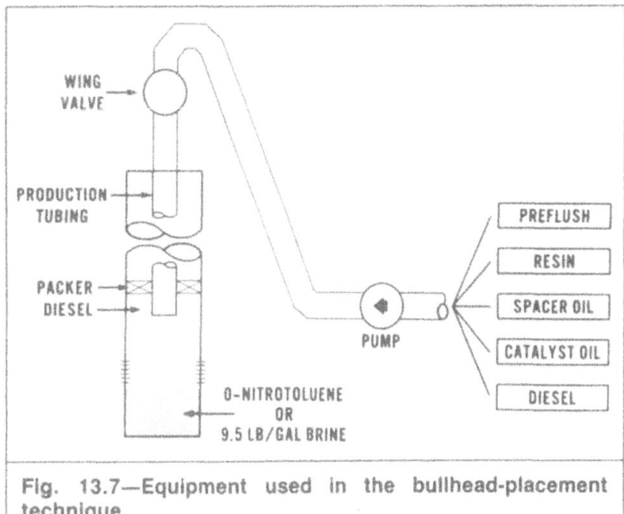

Fig. 13.7—Equipment used in the bullhead-placement technique.

Uses. Because no rig is required, the bullhead placement technique is more economical than the workstring methods and is not limited by rig availability or location. The one drawback is that the internal volume of the production tubing is frequently large compared with the volume of the chemicals pumped, leading to greater mixing of adjacent fluids. For this reason, small chemical volumes cannot be successfully bullheaded. In most instances, a minimum perforated interval of 6 ft should be treated to yield sufficient chemical volume for a bullhead placement procedure.

Bullheading is the preferred technique when a rig is not needed on location. Such instances include initial completions and recompletions where the zone is to be perforated and then immediately consolidated. If, however, a well is sanded up and requires a rig for cleanout, a workstring placement method would be preferred. One exception might be wells with low BHP's, which will not stand full of diesel. In these cases, bullheading might be preferable to running a workstring with a packer.

Restoring Well to Production

A well should be brought back to production slowly after a consolidation treatment. Unfortunately, few data are available to confirm the effect that initial production rates have on the ultimate success of sand-control techniques.

Swabbing and *gas lifting* are the two most common methods of restoring flow to the well. The gas-lift systems are designed to relieve the hydrostatic pressure slowly by unloading the kill fluid from the top down. By contrast, swabbing is potentially more detrimental. As the swab cup is pulled up, the hydrostatic head of fluid above the cup is instantaneously removed from the fluid below. If the well is swabbed too deeply or too rapidly, the instantaneous pressure from the reservoir into the well can become quite large. Presumably, the pressure drop could be sufficient to cause even a good consolidation to fail.

After consolidation, most wells will return to the productivity achieved before treatment. Wells that have been reperforated and effectively acidized before the consolidation treatment may produce at higher rates. Significant productivity loss caused by consolidation can be treated by either HCl acidizing or reperforating. However, if the problem is caused by poor reservoir quality, then little can be done to improve the production rate.

HCl Acidizing. Occasionally, productivity is difficult to restore after consolidation. Many reasons have been proposed to explain why productivity is lower even when injectivity was good while the consolidation treatment was being pumped. Two such reasons are perforation plugging with debris loosened from the casing wall by the powerful solvents in the resin systems and partially catalyzed resin draining down the casing wall and accumulating in the perforations.

Whatever the cause, acidizing with HCl after the consolidation resin has cured effectively enhances well productivity. HCl will not harm the consolidation strength once the resin has cured. The HCl can dissolve pieces of scale and rust in the perforations. It can also slightly attack the cement at each perforation, possibly enlarging the perforation tunnel and creating a new fluid flow path. Consolidations may be acidized with 50 to 100 gal/ft of 15% HCl. The acid is injected slowly and allowed to soak for a few hours before production is restored.*

Reperforating. If efforts to restore productivity with HCl have been unsuccessful, reperforation of the consolidated interval can be attempted. The same interval that was consolidated must be reperforated to avoid reaching sand that was not treated with resin. This requirement calls for good depth control and close attention by service company personnel.

The consolidated radius around the wellbore ranges from 1 to 4 ft, depending on reservoir stratification and the ability of the formation to accept consolidating fluids uniformly. This 1- to 4-ft radius is beyond the penetration distance of typical through-tubing perforating guns. Field experience has confirmed that intervals can be perforated without causing sand production.

Jet perforators penetrate because of pressure, not because of heat; hence, the jet will not melt or degrade the plastic. The rock pores should be open and available for flow as in any other perforating operation.

Summary

Field procedures for implementing plastic consolidation treatments have a significant impact on success. These include ensuring tubular integrity, having areas of high perforation inflow, and prepacking the perforations before the treatment. The completion interval should be injectivity tested before the treatment and acidized if insufficient injectivity exists. The treatment can be performed either through a workstring or bullheaded. In either case, the fluids must be segregated during pumping to avoid contamination and mixing downhole that can cause job failure.

Reference

1. Shaughnessy, C.M., Salathiel, W.M., and Penberthy, W.L. Jr.: "A New, Low-Viscosity, Epoxy Sand-Consolidation Process," *JPT* (Dec. 1978) 1805-12; *Trans.*, AIME, **265.**

SI Metric Conversion Factors

bbl	× 1.589 873	E−01	=	m^3
cp	× 1.0**	E+00	=	mPa·s
ft	× 3.048**	E−01	=	m
°F	(°F−32)/1.8		=	°C
gal	× 3.785 412	E−03	=	m^3
in.	× 2.54**	E+00	=	cm
lbm	× 4.535 924	E−01	=	kg
psi	× 6.894 757	E+00	=	kPa

*Unpublished procedures, Exxon Production Research Co., Houston (1970).
**Conversion factor is exact.

Chapter 14
Resin-Coated Gravel Packs

Introduction

Another technique to control formation sand involves consolidating a gravel pack with plastic resin. In this case, the gravel is sized according to the gravel-pack design techniques discussed in Chap. 4 and is consolidated after placement. Because the gravel is consolidated, no screen is required to hold the gravel in position, as **Fig. 14.1** illustrates.

Resin-coated gravel packs are pumped in slurries in much the same way as gravel prepacks. Before pumping, the gravel is coated with a resin, which can be either an epoxy,[1-3] a furan,[4,5] or a phenolic[6-10] resin. After placement in the well, the resins are cured to form a consolidated gravel mass. When the resin-coated gravel slurry is pumped into the well, all perforation tunnels should be filled so that sand can be controlled. The cured resin-coated gravel that remains in the well after the treatment is completed usually is removed before production is initiated.

Liquid Resin Coatings

To produce a liquid resin coating, gravel is first mixed with the transport fluid to form a slurry. A silane coupling agent can be added at this time to promote wetting and adhesion by the resin. Finally, the liquid resin is added to form a coating on the gravel. Because some resin diluent is extracted by the carrier fluid, the resin coating on the gravel becomes very viscous. Increased viscosity helps prevent resin removal during the placement operation.

Liquid resins are cured to a solid plastic with the aid of a catalyst. When the catalyst is added to the resin before coating the gravel, the resin is said to be catalyzed internally. When the gravel is initially coated with uncatalyzed resin and the catalyst is carried in a separate fluid pumped after the resin is outside the casing, the resin is catalyzed in situ.

Liquid resin systems involve chemicals similar to those of conventional plastic consolidations. Epoxy resins catalyzed internally with amines are used widely. Because they react very rapidly, resin coatings of liquid furan are usually catalyzed in situ by acids.

Solid Resin Coatings

A solid resin coating may be applied to gravel in two ways. It can be melted, mixed with the gravel, and allowed to cool; or it can be dissolved in a solvent and then mixed with the gravel. The solvent is allowed to evaporate, leaving the sand coated with resin. Because both methods require specialized processing equipment, the manufacturer applies the resin before shipment to the wellsite. Gravel coated with a phenolic resin is prepared by the solvent-type process.[6]

The resin coating is designed to be a solid at ambient temperatures when it is reasonably protected from moisture. The coated gravel is thus left dry and free-flowing, capable of being bagged and handled as ordinary gravel. The resin softens somewhat upon exposure to water or diesel. Thus, the sand grains will adhere to one another when immersed in liquid and compressed. Elevated temperatures (140°F) cure the resin to a hard plastic that binds the gravel into a hard, porous matrix. Steam injection or chemically induced heat may be considered if the reservoir temperature is too low to cure the plastic resin. Because the resin coating is thin, constituting only 5% of the product, permeability remains high.

When 30/50-U.S.-mesh, solid-resin-coated gravel is initially loaded with 100-psi compression and allowed to cure completely in either brine or diesel, the resulting compressive strength is about 2,500 psi. However, the compressive strength of the cured resin-coated gravel depends heavily on the compressive stress during curing. Tests have shown that samples cured at compressive stresses lower than 25 psi are much weaker than those cured at 100 psi. For maximum compressive strength, the curing stress should be high enough that the phenolic coating is penetrated and a grain-to-grain silica contact results.

Contacting the uncured coated gravel with brine or diesel containing 15% methanol or surfactants softens the resin coating and binds the grains together. When the gravel is pumped into a well with a circulating bottomhole temperature less than 140°F, overflushing with a softening agent is desirable. The softening-agent overflush will help fuse the grains together under hydraulic and overburden compression. This fusing can yield a two- to five-fold increase in cured strength, but the curing time for the resin is unchanged. As mentioned earlier, the temperature can be increased by pumping hot water or steam or by using chemicals that produce heat when they react with well fluids. It is not advisable to add a softening agent to the fluid used to pump coated gravel because the softened resin coating can stick to pumps, valves, and tubing.

Placement Procedures

Although each resin-coated gravel process has a recommended mixing and pumping procedure, most techniques involve the same placement fundamentals. Initially, the gravel is sized on the basis of the formation-sand-sieve analysis. This procedure commonly results in the selection of 20/40- or 40/60-U.S.-mesh gravel, but a larger size could be selected if conditions warrant.

The amount of slurry to be mixed depends on the length of the perforated interval. Generally, about 1 to 2 sacks of gravel/ft of perforated interval will be needed. In addition, sufficient gravel must be available to fill the rathole and a portion of the casing above the perforations.

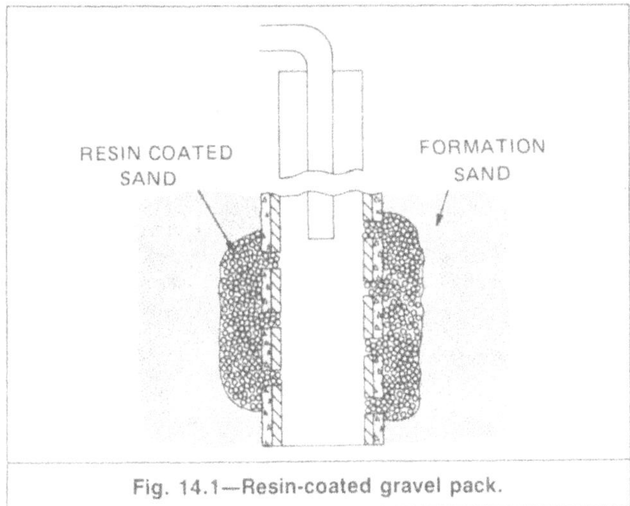

Fig. 14.1—Resin-coated gravel pack.

TABLE 14.1—COMMERCIAL RESIN-COATED GRAVEL SYSTEMS

Resin	Catalysis	Compressive Strength on 20/40-U.S.-Mesh Sand (psi)	Well Cleanout
Epoxy (liquid)	Internal	1,000 to 2,500	Drill
Furan (liquid)	In situ or internal	2,500	Wash or drill
Phenolic (solid)	Internal	2,500*	Wash or drill

*If cured at 100-psi confining stress. When cured at no confining stress, the compressive strength is considerably lower.

Once mixed, the slurry is injected into the well through a conventional workstring. Because excess sand must be removed from the well before production, a bullhead technique is not used. To minimize the mixing of slurry with other fluids in the casing, the end of the workstring is positioned 40 to 60 ft above the perforations. Preflush chemicals (if any) are pumped into the well, followed by the slurry. The slurry pumping rate is kept low to avoid fracturing the formation.

Slurry pumping continues until sandout. At this point, two procedures are possible. Internally catalyzed systems are usually allowed to cure in the sandout condition, with the gravel in the casing removed later with a drill or mill. In-situ-catalyzed systems may be reverse-circulated out of the wellbore and the sandpacking procedure repeated. After the final washout, a catalyst solution is injected to harden the resin.

Design Considerations

To obtain a successful treatment, gravel must be placed behind every perforation. Conceptually, this objective seems more difficult to meet than that of a consolidation treatment, which is to place a liquid resin behind each perforation. Coverage of plugged perforations from adjacent ones appears more likely for fluid flow than for sand transport. To achieve gravel placement through each perforation, perforation cleaning by washing or surging should be conducted on all wells before the resin-coated gravel is pumped. When applied to wells that have previously sanded up, a well cleanout may be all that is required before the resin-coated gravel is pumped through the perforations.

Complete interval coverage, however, is possible on the wellbore side of the casing. Specifically, most processes attempt to leave the perforation tunnels filled with resin-coated gravel. This condition certainly will result in initial sand-control success. But whether the gravel in this position can retain sufficient strength for long-term success is doubtful. Many of the early resin-coated gravel treatments were reported to have sanded out after only 1 or 2 sacks per well were placed. These treatments probably only filled the perforations. As one might expect, the lifetime of these early treatments was disappointing.

Resin-coated gravel is most successful as a remedial treatment on wells that have produced sand. Once some sand production has occurred, there is room outside the casing to allow 1 or more sacks of resin-coated gravel to be injected per foot of perforated interval. This 1-sack/ft guideline correlates well with treatment success. In newly perforated wells that have not produced, it is rarely possible to inject 1 sack/ft, and consequently initial job success has been lower. If a resin-coated gravel treatment is to be done on the initial completion, producing the well a short time before the sand is injected is preferable.

At present, there is considerable controversy over whether resin-coated gravel should be pumped at fracturing pressures or whether fracturing should be avoided. Treatments have been done with both techniques, but no clear direction can be derived from the field results. Proponents of fracturing believe that it will cause the gravel to be injected through more perforations, thereby improving coverage. Opponents believe that fracturing can cause the gravel to be conducted away from the well along a channel, leaving insufficient gravel at the perforations.

To overcome the problems associated with uneven gravel distribution, system formulators have incorporated extra resin to consolidate some of the formation sand with the gravel. The extra resin may be added to the initial resin gravel slurry or pumped as a separate step after the gravel is in place. If the resin exceeds the amount needed to coat the gravel and is incorporated into the initial slurry, resin will leak off from the slurry to the formation sand. This procedure increases the radius of the consolidation around the well, leading (one hopes) to an improved lifetime. Resin injected after placement of the resin-coated sand is usually done as a suspension in a transport fluid.

Commercial Systems

Table 14.1 is a compilation of the significant features of currently available, generic, resin-coated gravel systems.

Resin-coated gravel packs are routinely evaluated on the same basis as a plastic consolidation resin. Initial strength and permeability retention of the gravel are determined. Unfortunately, these tests represent the status of the resin-coated gravel before injection into the formation. No data are available for evaluating the pack after placement.

Because the gravel is resin-coated under controlled conditions at the surface, the choice of chemical ingredients is wider than it is in a conventional consolidation process. For this reason, a high-strength resin may be formulated with little difficulty. However, because there are fewer grain-to-grain contact points with 20/40-U.S.-mesh gravel than with formation sand, the consolidated strength of the 20/40-U.S.-mesh, resin-coated gravel is lower than in comparable plastic consolidation processes that consolidate formation sands. However, the initial compressive strength of most commercial resin-coated gravel is well above the 1,000-psi compressive strength needed to prevent sand production.

The permeability of either pure 20/40- or 40/60-U.S.-mesh gravel is much higher than that of typical formation sand. Therefore, calculations predict that productivity should not be impaired provided that the resin-coated gravel is sized correctly for the particular application. However, if there is sufficient mixing between the resin-coated gravel and formation sand during or after placement, well productivity will be adversely affected.

Summary

Resin-coated gravel packs use a combination of gravel packing and plastic consolidation. The gravel size is designed as recommended in Chap. 4. Gravel coated with a solid phenolic resin can be handled much the same way as liquids after the gravel is slurried with the carrier fluid. In either case, the resin-coated gravel must be placed in each perforation and allowed to cure into a consolidated mass because a screen is not used to hold the gravel in place.

References

1. Copeland, C.T. and McAuley, J.D.: "Controlling Sand With an Epoxy-Coated, High-Solids-Content Gravel Slurry," *JPT* (Nov. 1974) 1215–20.
2. Constien, V.G. and Mayer, M.H.: "What? No Screen? Gravel Packing with Water Carried Resin Coated Gravel," paper SPE 7003 presented at the 1978 SPE Formation Damage Control Symposium, Lafayette, Feb. 15–16.
3. Knapp, R.H., Planty, R., and Voiland, E.J.: "A Gravel-Coating, Aqueous, Epoxy Emulsion System for Water-Based Consolidated Gravel Packing: Development and Application," *JPT* (Nov. 1977) 1489–96; *Trans.*, AIME, **263.**
4. Young, B.M., Cook, J.G., and Donaldson, A.L. Jr.: "Here's a New Sand-Pack Method," *Oil & Gas J.* (March 10, 1969).
5. Sparlin, D.D.: "A New Sand Control Technique for Old Sand Problems," paper API 906-12-H presented at the 1967 API Spring Meeting of the Southwestern Dist., Div. of Production, Midland, March 15–17.
6. Cooke, C.E. Jr., Graham, J.W., and Muecke, T.W.: "Method for Treating Subterranean Formations," U.S. Patent No. 3,929,191 (1975).
7. Sinclair, A.R. and Graham, J. W.: "Super Sand Control—With Resin Coated Gravel," *Oil & Gas J.* (April 18, 1977) 58–60.
8. Sinclair, A.R. and Graham, J.W.: "An Effective Method of Sand Control," paper SPE 7004 presented at the 1978 SPE Formation Damage Control Symposium, Lafayette, Feb. 15–16.
9. Saunders, L.W. and McKinzie, H.L.: "Performance Review of Phenolic-Resin Gravel Packing," *JPT* (Feb. 1981) 221–28.
10. Underdown, D.R., Day, J.C., and Sparlin, D.D.: "A Plastic Precoated Gravel for Controlling Formation Sand," paper SPE 8801 presented at the 1980 SPE Formation Damage Control Symposium, Bakersfield, Jan. 28–29.

SI Metric Conversion Factors

ft	× 3.048*	E−01	=	m
psi	× 6.894 757	E+00	=	kPa

*Conversion factor is exact.

Chapter 15
Guidelines for Choosing a Sand-Control Method

Introduction

In deciding upon the most appropriate sand-control technique, the engineer must first be aware of the treatment options available (see **Fig. 15.1**) and then select one according to the geometric factors and completion requirements of the particular well.

Sand-Control Options

Once the decision has been made to perform a sand-control treatment, the question still remains as to what method to use: a chemical consolidation, a resin-coated gravel pack, or a gravel pack. This decision is based on wellbore and completion requirements and on economics.

Gravel Packing. Because of its versatility, gravel packing has been the most widely used mechanical sand-control technique. The screen length and opening size may be chosen to control many different sand sizes over any perforated interval length. The initial installation cost is relatively low but is tied directly to rig costs. If the completion interval is short and the well is on land, the cost will be much lower than for a similar situation offshore where a large floating drilling rig may be required. Gravel packs covering 20- to 200-ft intervals are common, and many treatments covering intervals of 500 ft or more have been performed.

Although gravel packing offers an economical method of controlling sand, it has several disadvantages. First, while initial installation is economical, a remedial treatment to replace a failed screen may involve an expensive fishing job. Second, in overpressured reservoirs that cannot be controlled with a calcium chloride brine, use of a special gravel-packing fluid is required. In these cases, the cost of the completion fluid could be large compared with screen and gravel costs, thereby changing the economics. Finally, productivity impairment caused by perforation tunnels being filled with sand and/or invasion of the gravel pack by formation fines is an important consideration. This problem is especially severe in small casings where very small screens must be used.

Bare Screens, Slotted Liners, or Prepacked Screens. The use of bare screens, slotted liners, or prepacked screens is also a possibility in certain situations; however, they usually should not be considered preferable to gravel packing in high-rate wells. They may have utility in certain wells where a short-term solution to a sand-control problem is needed and for economical or other reasons where a gravel pack is not a satisfactory solution.

Consolidation. The objective of chemical consolidation is to treat the formation in the immediate vicinity of the wellbore with a material that will bond the sand grains together at their points of contact. This treatment is accomplished by injecting liquid chemicals through the perforations and into the formation. These chemicals subsequently harden to form the bonding material (usually a plastic). For these treatments to be effective, two requirements must be met: the formation must be treated (consolidated) outside all perforations and the consolidated sand mass must remain permeable to well fluids.

Consolidation offers many advantages as a sand-control method. First, future workovers are simplified because no mechanical equipment is left in the wellbore. Therefore, an expensive fishing job is avoided in case of a sand-control failure. Second, because all the formation material is cemented together for several feet surrounding the wellbore, a productivity decline associated with the migration of fine particles toward the wellbore does not occur because the flow velocities are reduced at a treatment radius of 3 to 4 ft from the well. Finally, consolidation can be performed without a rig on wells that have not sanded up. This can be of particular benefit in initial completions and recompletions.

Offsetting these benefits are the two primary disadvantages of the consolidation technique—cost and coverage. The cost of a consolidation treatment is a minimum of $5,000/ft of perforated interval in 1992 U.S. dollars. At this price, the cost of consolidating a long interval (50 ft) becomes prohibitively expensive. However, if there is a reasonably good chance of success in a moderately long interval and the treatment can be performed without a workover rig, the consolidation treatment may be less expensive than a gravel pack. For short intervals (10 ft or less), the technique is competitive with other sand-control methods. The problems associated with coverage arise from injecting insufficient resin into low-permeability zones of highly stratified reservoirs. Because stratification is generally increased in longer intervals and because exact placement of the sequence of chemicals is required, consolidation success is reduced in perforated intervals longer than 15 ft. Hence, treatment of long intervals usually becomes impractical.

Resin-Coated Gravel. Like consolidation, the resin-coated gravel-pack technique relies on bonding gravel together to control sand migration. Instead of bonding formation sand, however, this technique consolidates gravel placed inside and immediately surrounding the perforation. The gravel (usually 40/60 or 20/40 U.S. mesh) is coated with a resin at the surface and then pumped into the well

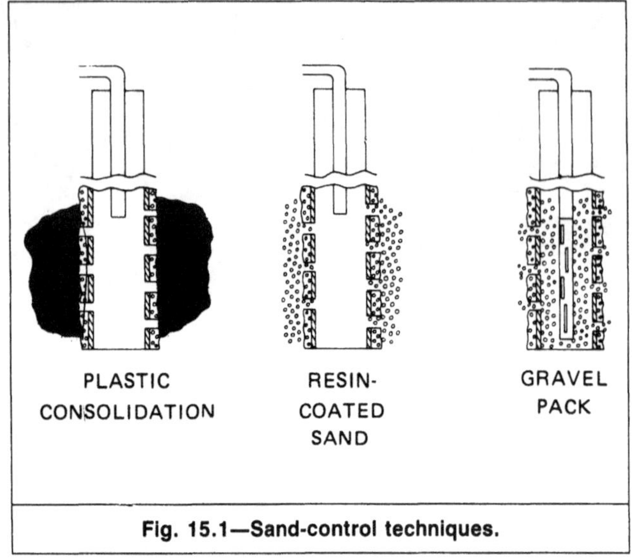

Fig. 15.1—Sand-control techniques.

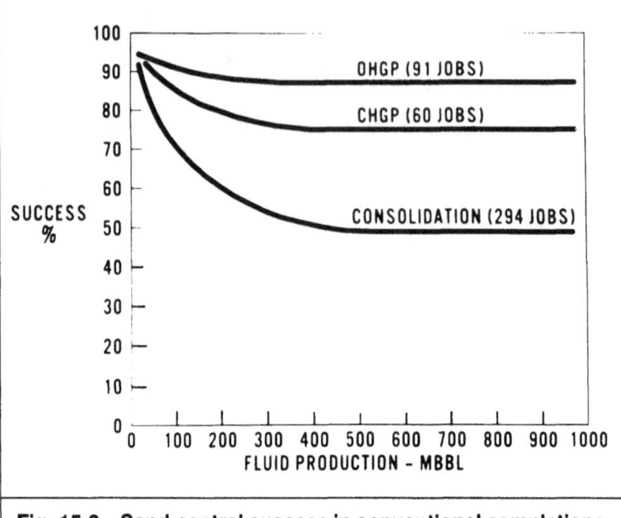

Fig. 15.2—Sand-control success in conventional completions.

as a slurry. This slurry is then squeezed through the perforations to fill a region behind the casing. After hardening, the consolidated gravel prevents formation sand from entering the wellbore. Excess resin-coated gravel is removed from inside the casing by either drilling or washing.

The primary advantage of this technique is reduced cost. Because of the limited amount of sand that is actually consolidated, the volume of resin needed is only 1 to 2 gal/ft of perforated interval. A resin-coated gravel treatment can be done for 25 to 50% of the cost of a conventional consolidation treatment or gravel pack, making it economically attractive in longer intervals.

A potential disadvantage of resin-coated gravel pack is its short lifetime. To be effective, a control method must stop sand production from every perforation. For a resin-coated gravel pack, this means forcing a significant amount of gravel through every perforation. The coverage problems appear to be even more severe than those associated with a consolidation treatment.

Because industry experience with resin-coated gravel packs has been limited, it is difficult to evaluate the performance of this treatment fully. In particular, sufficient data on resin-coated gravel packs are not available for assessing sand-control success in the various wellbore geometries and completion types.

Wellbore Configuration

Wellbore geometric factors (casing sizes and hole deviation) can play an important role in the effectiveness of the various sand-control methods.

Conventional Completions. Initial sand-control success in a conventional completion generally runs 90% or more; long-term success is less. Wells included in this category have 4½-in. (or larger) casings. They may be completed either open or cased hole.

Fig. 15.2 shows the success, in terms of cumulative fluid production, of openhole and cased-hole gravel packing, and plastic consolidation.* In this figure, success refers to the probability that a well will achieve a certain cumulative production provided that the reserves are sufficient. Obviously, not every cased-hole gravel pack in the study reached 800,000 bbl of cumulative production. But of those having the opportunity to produce 800,000 bbl of fluid (sufficient reserves), only 25% failed and 75% were still controlling sand.

The implication of Fig. 15.2 is fairly clear: openhole gravel packing is the most successful sand-control technique. Unfortunately, only a few wells usually meet the long-life, productivity, and minimum recompletion requirements necessary for successful openhole production. Therefore, the choice between a cased gravel pack or a plastic consolidation must frequently be made. Both methods have a high initial success rate, with most of their failures occurring early in the wells' production lives. In the long run, gravel packing is more successful than plastic consolidation in conventional completions.

*Unpublished Exxon Co. U.S.A. report, Houston (1976).

Fig. 15.3 shows another aspect of gravel packing in conventional cased-hole completions as production rate vs. time after gravel packing.[1] Upon gravel packing, well productivity was cut by almost 50% because gravel filled the perforation tunnels. The consequences of failure, including those factors that influence productivity after a gravel-pack recompletion, are clearly demonstrated in the field performance depicted in Fig. 15.3. In this particular case, productivity might have remained unaffected by the gravel pack if a simple inflow-performance analysis had been made. If the well had been reperforated with a higher shot density or with large-diameter perforations, gravel packing would not have impaired productivity. The projected well performance after reperforating, also illustrated in Fig. 15.3, clearly shows the benefits of these practices.

Unlike gravel packing, plastic consolidation does not involve filling the perforation tunnels with gravel. Prepacking the perforations may be necessary, however, if the treatment was conducted as a workover and a significant amount of formation sand had previously been produced. Because the perforation tunnels are left open after a consolidation treatment, the productivity loss is believed to be 15 to 20%. Because the formation fines are consolidated for several feet around the wellbore, the productivity decline with time is less with a plastic consolidation treatment than with a gravel pack.

Tubingless Completions. Fig. 15.4 shows the success data for plastic consolidation and cased-hole gravel packs in tubingless completions.* This example shows the results of many different consolidation treatments performed over a period of time. The results are penalized by many inferior treatments performed in early development stages but not in use today. The results of more recent sand-consolidation technology are superior to these data. The success of tubingless gravel packs is definitely lower here than in conventional completions. No tubingless wells were completed open hole.

Note that the initial gravel-pack success is quite good but declines rapidly. Gravel-pack success is reduced in tubingless completions for several reasons. First, the screen placed inside 2⅞- or 3½-in. tubing is 1 to 1¼-in. in diameter. The slot area of such a screen is small and therefore easily plugged by even a small amount of fine material from the formation. Second, less than 1 in. of gravel separates the screen from the perforations, even if the screen is perfectly centralized. This greatly increases the chances that formation sand will reach the screen. A sand-control failure usually results in this circumstance because the screens are not designed to bridge off formation sand. Third, tubingless wells were ordinarily perforated at 4 shots/ft rather than at the 8 to 12 shots/ft for a conventional completion. Also, the only perforating guns available are those that shoot small-diameter holes that restrict productivity, which also

*Unpublished Exxon Co. U.S.A. report, Houston (1976).

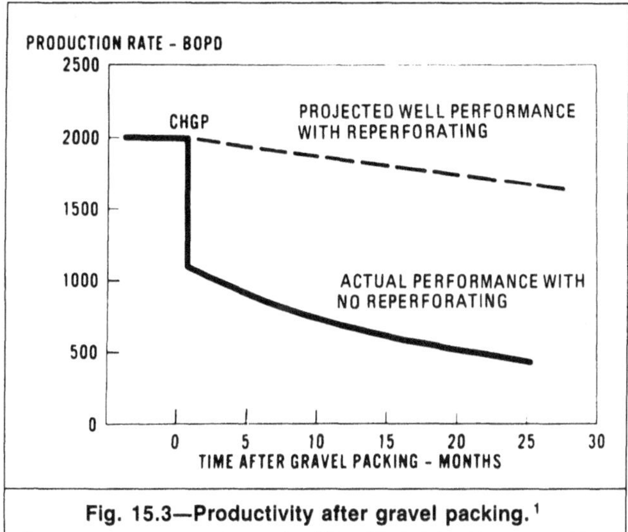

Fig. 15.3—Productivity after gravel packing.[1]

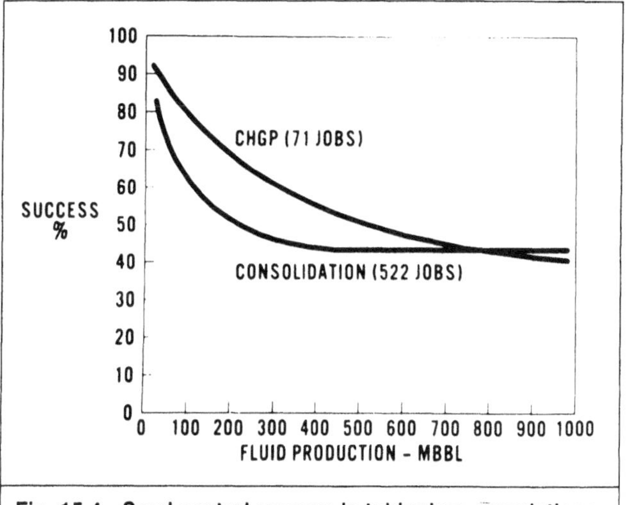

Fig. 15.4—Sand-control success in tubingless completions.

tends to reduce completion life. When a reduced perforation area becomes filled with sand, well productivity can dwindle dramatically. Finally, the majority of these wells were gravel packed in the early 1970's, when such procedures as perforation cleaning and using clean fluid were not as refined as they are now.

Plastic consolidation in tubingless completions is about 5% less successful than in conventional completions. A frequently cited reason is the difficulty encountered in achieving a good primary cement job in multiple tubingless wells. A channel in the primary cement may prevent uniform formation-sand consolidation outside every perforation, which tends to reduce both initial and long-term job success (see **Fig. 15.5**).

Well Deviation. Completion success in wells with different deviation angles has not been studied extensively. In gravel packs, the basic problem is completely filling the annulus between the casing and screen with gravel. For wells inclined more than 45° from vertical, totally filling the annulus with gravel may not be possible unless proper placement designs are implemented. However, special tools and placement techniques designed to overcome this problem have been developed.

Neither field nor laboratory results indicate that well deviation significantly affects plastic consolidation. However, no study has specifically isolated deviation angle as a factor in success or failure.

Reservoir Properties

The properties of reservoirs where sand must be controlled will influence the type of sand-control method used. Interval length and sand quality are often the deciding factors, but reservoir temperature, pressure, and fluid should also be considered.

Interval Length. Economic considerations often require that long intervals be gravel packed. With interval lengths less than 20 ft, however, such decisions must also be based on success and productivity considerations. Earlier sand-control success curves did not differentiate between long and short interval lengths. Gravel-packing data were collected primarily on wells with long production intervals; consolidation data referred almost exclusively to short zones. Long intervals have an advantage in a lifetime study concerned only with cumulative production because their total flow rates are usually higher than those in short intervals.

To eliminate distortion of conclusions relative to interval length, sand-control success comparisons have been made of intervals of similar length. Fig. 15.5 compares gravel-packing and consolidation success rates in intervals having <12-ft perforations.* Gravel packing still has the highest success rate, followed by plastic consolidation and finally by tubingless gravel packing.

*Unpublished Exxon Co. U.S.A. report, Houston (1976).

Sand Quality. "Sand quality" refers to the nonsilica content and grain-size distribution of sand. In general, formation sands are composed of sand and shale. Sand is primarily SiO_2. Shales contain, in various proportions, clay minerals (illite, montmorillonite, and kaolinite), silt, carbonate, and other nonclay materials. Silt is a very-fine-grained material that is predominantly quartz but that may also include feldspar, calcite, and other minerals.

A good-quality sand has a narrow grain-size distribution and a low nonsilica content (i.e., 5 to 15%). The permeability of such sand is generally high (1 darcy or greater). A poor-quality sand, on the other hand, generally has a wide particle-size distribution, an appreciable fraction of which will pass through a 400-U.S.-mesh screen. The nonsilica content of these sands can be as high as 50%, and permeability may be reduced to 100 md or less.

Information on sand quality comes from many sources. Direct evidence includes rubber-sleeve cores, sidewall cores, and a completion log constructed from cuttings and the like as the well is drilled. Indirect evidence of formation strength can be obtained from the drilling-time log, the caliper log, and a combination (sonic/density) logging tool called the mechanical-properties log. Clay content/shaliness can be estimated with increasing precision from well logs, dual-log crossplots, and multiple-tool-computed logs.

In general, as sand quality deteriorates, the success of all sand-control techniques also declines. Sand consolidating chemicals exhibit poorer wetting and adhesion performance with nonsilica materials than with sand. This can lead to a low-strength consolidation incapable of withstanding the stress imposed during production. Gravel packs are usually too coarse to bridge off the very fine formation particles successfully. These particles soon invade the pack and either plug or erode the screen. Sand consolidation success tends

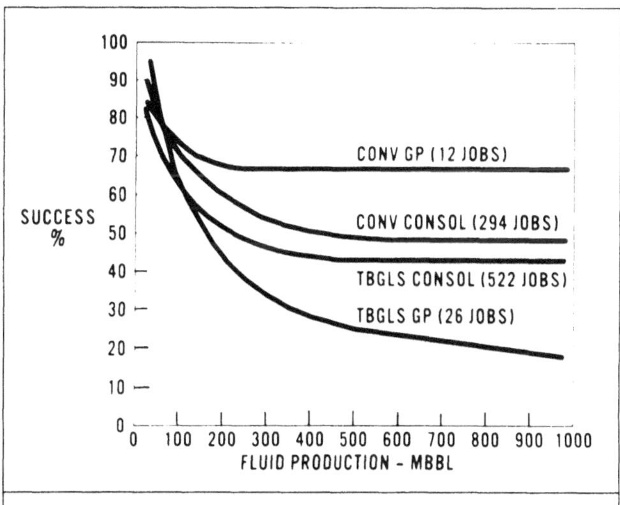

Fig. 15.5—Sand-control success in intervals less than 12 ft.

to decline more rapidly with deteriorating sand quality than gravel packing does for several reasons. The lower the permeability, the more difficult it is to place the plastic and chemicals under matrix conditions. Also, the plastic tends to occupy a greater fraction of the pore space in low-permeability than in high-permeability sands.

Reservoir Temperature. As a rule, reservoir temperature does not play an important role in the selection of a sand-control technique. With few exceptions, all techniques can be used in reservoirs with a static bottomhole temperature ranging from 120° to 200°F, a range that probably encompasses 95% of all sand-producing wells. Gravel packing is applicable at temperatures both above and below this temperature range, but many consolidation and resin-coated gravel systems are not. The phenolic-resin-coated sand, however, is stable up to about 400°F.

Because consolidation involves a chemical reaction, the hardening process may occur too slowly if the temperature is low. To combat this lag, special reaction accelerators and catalysts can extend the useful temperature range of some processes down to 75°F. At extremely elevated temperatures, many bonding agents become soft, causing the consolidated sand to lose strength. Once again, special resin formulas are available to extend the useful range of some systems to 250°F.

Reservoir Pressure. Like reservoir temperature, reservoir pressure is not usually an important factor in the selection of a technique. Workover fluids are commonly available to control formation gradients from 0.30 to 0.60 psi/ft. Special completion fluids may be required in abnormally pressured reservoirs.

Reservoir Fluid. The type of fluid (oil, gas, or water) produced from the reservoir should be considered during technique selection. Fine particles entrained in a high-velocity gas stream may severely erode mechanical equipment. The erosion problem is even worse when gas is in turbulent flow, which it may well be (depending on rate) as it moves through the gravel pack and screen. Screen erosion has been observed in those sections of the completion where it is exposed.

Summary

Sand-control alternatives include gravel packing, sand consolidation, and resin-coated-gravel completion and production practices. Which technique is chosen depends on well and reservoir conditions, costs, and the treatment that will provide the maximum sustained productivity.

Reference

1. Williams, B.B., Elliot, L.S., and Weaver, R. H.: "Productivity of Inside Casing Gravel-Packed Completions," *JPT* (April 1972) 419-24.

SI Metric Conversion Factors

bbl	× 1.589 873	E−01	=	m^3
ft	× 3.048*	E−01	=	m
°F	(°F−32)/1.8		=	°C
gal	× 3.785 412	E−03	=	m^3
in.	× 2.54*	E+00	=	cm
psi	× 6.894 757	E+00	=	kPa

*Conversion factor is exact.

Author Index

Subject Index

P

T

U

V

W

X

Y

Z